P9-DES-833

ALSO BY AMIR ALEXANDER

Infinitesimal:
How a Dangerous Mathematical Theory Shaped the Modern World

Duel at Dawn:
Heroes, Martyrs, and the Rise of Modern Mathematics

Geometrical Landscapes:
The Voyages of Discovery and the Transformation of Mathematical Practice

PROOF!

PROOF!

HOW THE WORLD BECAME GEOMETRICAL

≡

AMIR ALEXANDER

SCIENTIFIC AMERICAN / FARRAR, STRAUS AND GIROUX

New York

Scientific American / Farrar, Straus and Giroux
120 Broadway, New York 10271

Printed in the United States of America
First edition, 2019

Library of Congress Cataloging-in-Publication Data
Names: Alexander, Amir R., author.
Title: Proof! : how the world became geometrical / Amir Alexander.
Description: First edition. | New York : Scientific American : Farrar, Straus and Giroux, 2019. | Includes bibliographical references and index.
Identifiers: LCCN 2019000250 | ISBN 9780374254902 (hardcover).
Subjects: LCSH: Geometry—History. | Geometry—Popular works.
Classification: LCC QA443.5 .A44 2019 | DDC 516.009—dc23
LC record available at https://lccn.loc.gov/2019000250

Designed by Richard Oriolo

Our books may be purchased in bulk for promotional, educational, or business use. Please contact your local bookseller or the Macmillan Corporate and Premium Sales Department at 1-800-221-7945, extension 5442, or by e-mail at MacmillanSpecialMarkets@macmillan.com.

www.fsgbooks.com • books.scientificamerican.com
www.twitter.com/fsgbooks • www.facebook.com/fsgbooks

1 3 5 7 9 10 8 6 4 2

To the memory of my mother,
Esther Alexander,
1929–2005

It is marvelous enough that man is capable at all to reach such a degree of certainty and purity in pure thinking as the Greeks showed us for the first time to be possible in geometry.

—ALBERT EINSTEIN

CONTENTS

PROOF!

INTRODUCTION

The Superintendent's Crime

The sun was setting in the summer sky of August 17, 1661, when the young king of France descended from his gilded carriage onto the front steps of the château of Vaux-le-Vicomte. Only a few years in the making, and still under construction, the château was the pride and glory of its owner, Nicolas Fouquet, the king's superintendent of finances. At Vaux, he had built his dream house and garden, a magical land of canals and fountains, woods and fruit groves, open vistas and hidden springs. Sparing no expense in his preparation for the royal visit, Fouquet was determined to dazzle his guest with a celebration worthy of a king.

Fouquet first led the king and his retinue on a tour of the elegant

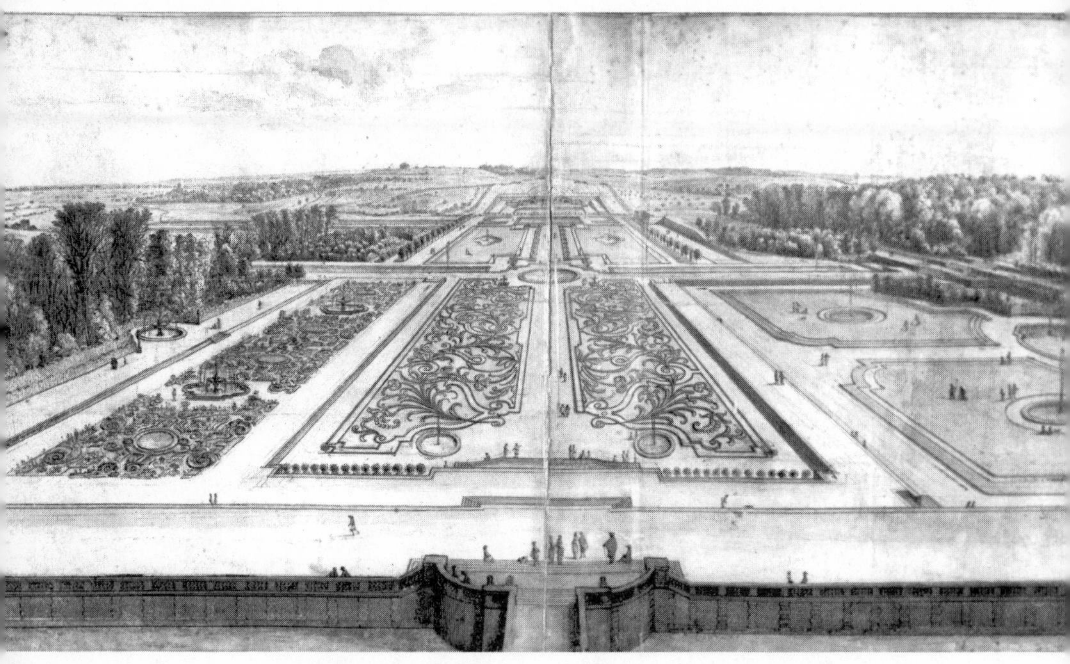

FIGURE 1: The gardens of Vaux-le-Vicomte today, pen and ink and watercolor by Israel Silvestre the Younger (1621–1691)

new château, designed by his favorite architect, Louis Le Vau. He guided them through the main hallway, decorated by France's leading painter, Charles Le Brun, then showed them on to the gardens behind it, created by the inimitable André Le Nôtre. There, they walked through the broad central avenue that stretched southward from the center of the château to the garden's farthest end, dividing it into two near-identical halves. They passed between rectangular parterres, known as the Embroidery for their exquisitely-gardened patterns, and gazed into the placid water circle between the two straight branches of the central canal, leading east and west. Proceeding between the symmetrical pools of Triton they paused to meditate before the great square mirror of water and finally settled down in front of a semicircular grotto. There they were treated to a performance of *The Bores*, a new comedy-ballet Molière had written for the occasion. A sumptuous meal served to each member of the royal court and the five thousand household guards was followed by a display of fireworks. They shot up from the château's central dome and lit up the sky, then slowly descended on the gardens like midnight suns.

The next morning Fouquet saw off the king and his retinue, still basking in the glow of royal approval. But as the king settled himself in the royal carriage on his way to the royal castle of Fontainebleau, he turned to his aging mother and remarked with disdain: "Ah, Madame, should we not make these people disgorge all of that?"[1]

Louis XIV was only twenty-three years old at the time, but already a man of his word. Three weeks after the festivities at Vaux he invited Fouquet to join him on a royal trip to Nantes, and there he set his trap. While showing every sign of favor in an audience with the superintendent, he secretly summoned the Comte d'Artagnan, captain of the King's Musketeers, to the adjacent room. D'Artagnan, the historical inspiration for Alexandre Dumas's famous hero, had known Fouquet for years and had benefited from his largesse, but he did not hesitate. At a sign from the king the dashing musketeer sprang forward and placed his former friend under arrest.

Soon after, Fouquet was put on trial for despoiling the royal treasury. His personal popularity and friends in high places helped sway the judges sufficiently to spare his life and sentence him to permanent exile. But the king would not be denied: by royal decree he "commuted" Fouquet's sentence from exile to life imprisonment and kept him locked up in an alpine fortress for the remainder of his days. He died a forgotten man in 1680, without ever setting eyes on his beloved estate again.

How had the loyal Fouquet, who had known Louis since childhood, incurred the king's wrath? Partly, no doubt, it was his success that made him a marked man. Fabulously wealthy, the owner of a private fleet of armed merchant vessels, and master of the strategic island of Belle-Isle, Fouquet was an obvious target for a young king looking to assert his authority. But power politics alone cannot explain Louis's vengefulness. After all, the king had been lenient with the leaders of the Fronde—an uprising that marred the early years of his reign—and he forgave the wealthy grandees of the Parlement of Paris for their key role in it; he tolerated the questionable loyalty of the provincial nobility, choosing to draw them to his court rather than bring them to heel; and he

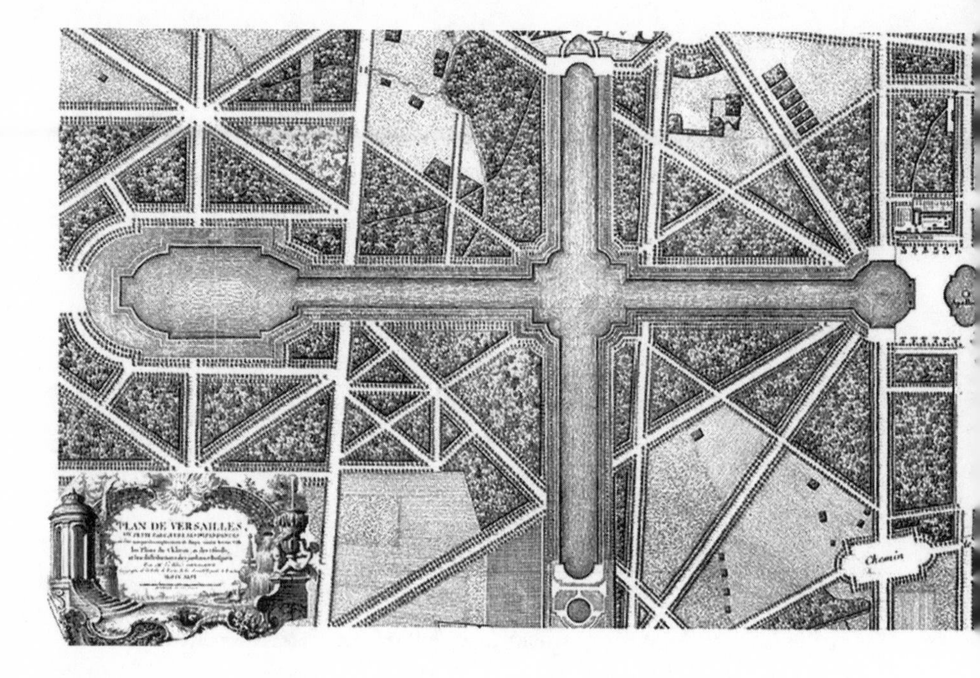

entirely forgave his ablest general, the Grand Condé, who had gone so far as to take up arms against his sovereign. But for Fouquet there was no mercy. Though he professed nothing but love and loyalty to his king, he was condemned by his implacable monarch to a cold and lonely prison cell from the day of his arrest to his death nineteen years later—long after he could possibly have been considered a threat.

Fouquet's offense was clearly unforgivable, but what exactly was it? To understand this we must return to the scene of his crime, the beautiful château of Vaux-le-Vicomte. Shortly after Fouquet's arrest an army of looters descended on the grounds like a swarm of locusts. In the gardens they uprooted trees, shrubs, and flower beds, pulled out ancient statues and bubbling fountains. From the château they took paintings, sculptures, and gilded chandeliers. But the raiders were not brigands from the countryside: they were the king's men, sent to take down all the objects that had made Vaux the enchanting palace that it was. They carefully packed the objects into crates and hauled them away to a marshy town where Louis was intent on building his own dream palace. It was called Versailles.

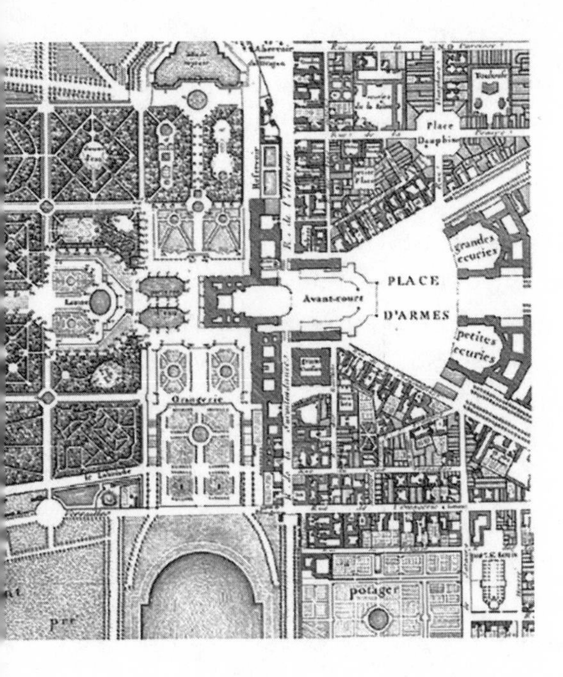

FIGURE 2: **Abbot Delagrive, ground plan of the gardens of Versailles, 1746**

Nor were fountains and sculptures the only things Louis transported from Vaux to Versailles; he brought the people as well. With Fouquet safely out of the way, the king immediately enlisted Le Vau to design his new palace. Charles Le Brun, the painter who had decorated the interiors of Vaux, now had the same job in Versailles, and the literati who had lent brilliance to Fouquet's court were packed off to Versailles to perform the same duties for the king. But Louis's greatest acquisition from his fallen courtier was not a writer, a painter, or even an architect. It was the gardener, André Le Nôtre. The king appointed him Chief Architect of the gardens of Versailles, and immediately set him to work recreating Fouquet's gardens on an immensely grander scale.

IT WAS AT VERSAILLES THAT the true nature of the superintendent's crime was revealed. It was not the superintendent himself who had provoked Louis's wrath, for Fouquet could perhaps have been forgiven: it was, rather, his sparkling geometrical estate that the king could not tolerate. It was Vaux-le-Vicomte that had become the king's obsession and drove him not only to despoil it but to co-opt it, and to do so on a scale that would dwarf the original and erase it from memory. And so, in the end, it was not Fouquet's wealth, or his private fleet, or his hasty decision to fortify his island of Belle-Isle that did him in; it was his geometrical garden.

The notion that a garden, even a beautiful one, would loom so large in the mind of a great monarch might surprise us. It's not, after all, as if Louis XIV had time and resources to spare: in the same years in which he broke Fouquet and embarked on the grand redesign of Versailles, Louis was also busy consolidating his personal rule over recalcitrant courtiers and ministers, as well as establishing his authority over provincial nobles, separatist Protestants, and city *parlements*, aristocratic bodies that served as the highest courts in the land and were charged with ratifying the king's decrees. Repeatedly, he sent his army to suppress uprisings in the provinces and launched aggressive wars against France's neighbors. Was this really a time to spend scarce resources on a vanity project like the gardens of Versailles? Was not the king's obsession with Vaux-le-Vicomte getting the better of his judgment? Jean-Baptiste Colbert (1619–1683), for one, thought that it was. As Fouquet's successor in charge of the kingdom's finances, he never ceased complaining to the king about the bottomless money pit he was building at Versailles.

Yet there is much to be said for Louis's position, for Vaux—and consequently Versailles—was not merely a garden but a geometrical one. And geometry, as this book will show, was not the obscure mathematical field it is today but a discipline of towering authority: it was the science of rational order itself, and its theorems, rigorous and irrefutable, stood for a deep unchallengeable order that pervaded the universe. For Louis, for Fouquet, and for their contemporaries, geometry was everywhere and structured everything—from physical nature to human society, the state, and the world. It stood for perfect order, hierarchy, and harmony, governed by the incontestable laws of reason. It stood, in other words, for a complete vision of the modern world that was then coming into being. And so when Fouquet created his own geometrical microcosm at Vaux, he was not only boasting of his fabulous wealth and cultured tastes: he was putting his own spin on the proper order of the world and the kingdom, and challenging his own circumscribed place within them. It was, sadly for Fouquet, a colossal miscalculation. A geometrical challenge, he soon found out, was a challenge the king could not ignore.

How and why did geometry acquire its exalted place in the Western tradition? How did it become an emblem of universal order and a tool of political power? How did it shape our past, and in what ways does it still mold our present? All of these questions will be addressed in this book. But before we begin we need to travel back two and a half millennia, and land on a sunny Mediterranean shore. For it was there, somewhere in the Greek-speaking world around 400 B.C.E., that the very first geometrical proofs were discovered and written down. And the world, it is fair to say, was never the same again.

Euclid's Kingdom

It is surely no coincidence that among all the peoples of the ancient world, it was the Greeks who produced the first geometrical proofs. This is not because the Greeks were the only nation to develop an impressive mathematical tradition. Far from it: the Egyptians made use of arithmetic for administrative and timekeeping purposes, and the construction of monumental edifices such as the pyramids necessarily required knowledge of practical geometry. Babylonian priests, who conducted the earliest systematic astronomical observations, developed sophisticated methods of calculation that some historians consider a form of algebra. A lively tradition of practical mathematics flourished in China, and Mayan priests in the Yucatan made complex calendrical calculations. The Hindu mathematical tradition, meanwhile, is the most likely originator of the seminal concept of "zero."

Yet despite their impressive originality and technical acumen, no Egyptian, Babylonian, or any other mathematician of the ancient world ever produced a mathematical proof. And little wonder: mathematics, for them, was a practical tool—a means to calculate such things as taxes, storage capacity, the positions of the planets, or how to erect a building that will not collapse. All of these require some common sense, mathematical acumen, and a great deal of experience. But none require a proof. For a proof is not about discovering any particular useful or interesting result: it is about establishing absolute and irrevocable truth. And while others might seek true knowledge in tradition, ancient authority, or divine revelation, it was only the Greeks who believed that the truth could be discovered by human reason alone. Consequently, it was only the Greeks who thought that absolute truth could be discovered through that most systematic and rigorous form of human reasoning—mathematics.

We do not know exactly when or where the first proof was written down, though it was almost certainly somewhere in the Greek-speaking cities that dotted the shores of the Mediterranean Sea in the fifth century B.C.E. Nor do we know what the first mathematical proof was, or the name of the anonymous genius who conceived it. It was, in all likelihood, a simple argument about lines and angles, which would be familiar to any beginning student of geometry today. And yet the implications of the discovery were momentous: for the very first time a truth about the world had been demonstrated with absolute logical certainty, making it impervious to challenges, counterarguments, or doubt. No one before had ever done so. No one, in fact, had ever thought of doing so, or imagined it was possible. And yet here it was—an irrefutable universal truth, arrived at simply by the disciplined application of human reason. It was a stunning and unprecedented revelation. And whereas other scientific and technological discoveries had been made over and over again, by different peoples in different times and places, this was not the case for the geometrical proof: it was a discovery made only once in all of human history.

Although the first proofs passed unremarked by contemporaries, they nevertheless circulated among geometers in the Greek world, who used them to produce additional proofs and mathematical results. Most of the work of these fifth-century practitioners has been lost to us, but what we do know from later sources is impressively sophisticated. Indeed, some of the classical geometrical problems of antiquity have their origins in this early period: One, known as the Delian problem, concerned the doubling of the cube (i.e., constructing a cube whose volume is double that of a given cube). Another was "squaring the circle," or constructing a square whose area was equal to that of a given circle. Neither question was fully solved since—as was proved more than two millennia later—they are not in fact solvable by the methods of Greek geometry. Work on them nevertheless produced impressive results.

The greatest of the late fifth-century geometers, Hippocrates of Chios, for instance, discovered that the doubling of the cube is equivalent to finding two mean proportionals between a magnitude a and its double, $2a$. This means that if $a/x = x/y = y/2a$ then $x^3 = 2a^3$, and consequently a cube with the side x has a volume twice as large as a cube with the side a. In another tour de force Hippocrates demonstrated that the area of a crescent-shaped "lune" is equal to the area of a right-angled triangle whose hypotenuse is equal to the lune's diameter.[2]

The complexity and sophistication of early Greek geometry are in themselves remarkable. But even more so is the fact that unlike mathematicians from Babylon to Mexico, Greek geometers were not looking to use their results for particular ends, or at least not exclusively so. They were, rather, looking to discover new geometrical relations simply because they were true. In other words, they were seeking truth for the sake of truth. It was the philosopher Plato (429–347 B.C.E.) who gave the most enduring expression to this attitude, arguing in his dialogues that geometry provided unparalleled access to a world of universal truths. According to Plato the physical world we know through our senses is but a pale and transient shadow of the "real" world, composed of eternal, unchanging "forms." Geometry, he wrote, which begins with self-evident assumptions and proceeds through strict logical reasoning, discards everything that is erroneous, unessential, and transitory, leaving us with truths that approach the eternal forms. Little wonder that—if tradition be credited—Plato ordered an inscription carved above the entrance to his Academy in Athens: "Let no one ignorant of geometry enter here."

For all his admiration of geometry, however, Plato was not himself a geometer. He celebrated the accumulation of geometrical knowledge, praised its method and proofs, and argued that they laid out a path to universal truths. But while he believed that geometry offered a glimpse of the world of the forms, he had little to say about what that perfect geometrical world looked like. The task of actually mapping out the geometrical world was therefore left to one who possessed both a deep technical understanding of the geometrical practices of his day and a broad knowledge of the results produced by the preceding generations of geometers. This man was Euclid, a professional geometer and scholar in the great Museum of Alexandria around the year 300 B.C.E.

Euclid is likely the most influential mathematician who ever lived, yet he may never have produced a single original mathematical result. No creative mathematical technique, or ingenious construct, is associated with his name: there is no "Euclid's sieve" on a par with the method for discovering prime numbers named after Eratosthenes of Cyrene, or "Euclidean spiral" like the one made famous by Archimedes.[3] For Euclid was not looking to discover novel mathematical results: his fellow geometers were already producing plenty of those, and had been for more than a century. Instead, he set his sights much higher: he wanted, rather, to create a complete mathematical world. And that is precisely what he accomplished in the *Elements*.[4]

To mathematicians of Euclid's day, the *Elements* contained few if any new insights that would help them advance the frontiers of knowledge. The theorems about parallels, triangles, and circles that could be found there were for the most part well known, and geometers applied them routinely in their work.[5] Euclid, however, gathered all those familiar but disparate results, brought them together, and then turned them into a single consistent logical system.

The *Elements* begins with a series of postulates and common notions, so obvious and self-evident that our minds compel us to accept them as true. "All right angles are equal to one another," states Postulate 4; "If equals be added to equals, the wholes are equal," states Common Notion 2; "The whole is greater than its part" is Common Notion 5.[6] From these postulates, Euclid moves on to prove ever more complex

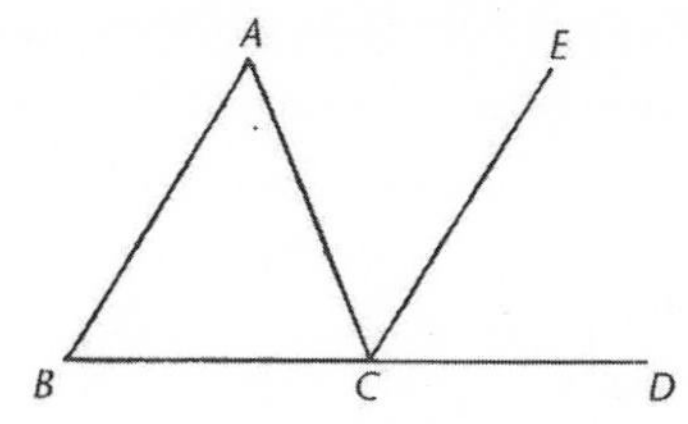

FIGURE 3: **Euclid, *Elements*, Book I, Proposition 32**[7]

truths: that the sum of the squares of the legs of a right triangle is equal to the square of the hypotenuse (*Elements* I.47); that a line touching a circle at a single point is perpendicular to a line from that point to the circle's center (*Elements* III.18); that in a circle, the angle in a semicircle is a right angle (*Elements* III.31); and much, much more.

Consider for example Euclid's proof in *Elements* I.32, that the sum of the angles of any triangle is equal to two right angles, or what we commonly call 180 degrees. He begins by considering the triangle *ABC*, and extending the side *BC* to the point *D*. He then proceeds:

First, he draws a line *CE* (through the point *C*) parallel to *AB*. That it is possible to draw such a line, and the means to do so, are demonstrated in Proposition I.31.

Then, since the lines *AB* and *CE* are parallel, and the line *AC* intersects them, the angles *BAC* and *ACE* are equal. This is true because Proposition I.29 had demonstrated that when a line intersects with parallels, alternate angles (in this case *BAC* and *ACE*) are equal.

Furthermore, since the lines *AB* and *CE* are parallel, and the line *BD* intersects them, the angles *ABC* and *ECD* are also equal. Once again this is based on Proposition I.29, which proved that when a line (in this case *BD*) intersects with parallels, the interior and opposite angles (in this case *ABC* and *ECD*) are equal.

Consequently, the angle *ACD*—which is the sum of *ACE* and *ECD*—is equal to the sum of the triangle's interior angles *BAC* and *ABC*.

Adding the angle *ACB*, we get that the sum of the triangle's three angles (*BAC*, *ABC*, and *ACB*) is equal to the sum of the adjacent angles *ACD* and *ACB*.

But the adjacent angles *ACD* and *ACB* are created by the line *AC* intersecting the line *BD*, and Proposition I.13 had demonstrated that the sum of such angles is always equal to two right angles.

Consequently the sum of the angles of the triangle is equal to two right angles.

Q.E.D.[8]

THE DEMONSTRATION IS A MICROCOSM of the logical relations that make up the *Elements*. The proof of Proposition I.32 is based on a series of earlier proofs about lines, parallels, and angles (I.13, I.29, I.31), which in turn are based on earlier propositions, all the way down to simple, self-evident postulates and common notions. Meanwhile the result of I.32, that the sum of the angles of a triangle is equal to two right angles, is itself used to prove ever more complex results. In this manner every proposition in the *Elements* is supported by other propositions, and in turn supports still others, branching out in all directions. Taken together they form a vast web of interlocking truths, in which each statement is related to all others and kept in an unchanging logical relationship to them. The end result is a complete world of mathematical truths, founded on indisputable postulates and proved through impeccable logic. It is, without a doubt, one of the greatest creations of human history: in the *Elements*, the dream of a rational and eternal world of absolute truth, so elegantly evoked by Plato, becomes reality.

And what a world it is! Because it is founded on indisputable first principles, and produced through flawless chains of deduction, it is entirely rational, and everything in it is irrefutably true. Because both the assumptions and the reasoning are valid everywhere and always, all objects in this world are universal and timeless. Because all theorems (i.e., proven propositions) in the *Elements* are logically interlocked, and exist in a fixed relationship to one another, this world is perfectly ordered, and nothing within it can be altered. Finally, because each theorem is founded on simpler ones, layer by layer, down to the original postulates, this world is hierarchical: each theorem has its unalterable place in a single great hierarchy of Truth. When we read the *Elements* we are not just learning a succession of increasingly complex geometrical truths; much more, we are being given a glimpse of an ideal world that exists unchanging through all time and space, perfectly rational, perfectly ordered, and hierarchical through and through.

The Imperfect World

But where does this world exist? Since it is universal and timeless, it would seem to follow that it is all around us, and that our own world is—or at least should be—perfectly ordered and hierarchical. There is, in fact, some evidence that geometers and philosophers came to precisely this conclusion: Plato, for instance, who admired the harmony and order of geometry, also detested the endemic chaos of political life in democratic Athens. In the *Republic*, in the same dialogue in which he extolled the virtues of geometry, he also advocated for a static political order in which each citizen is assigned a place for life in a rigid class hierarchy. Reason itself, Plato argued, determined that this was the only correct way to order the state. And while ancient geometers refrained from bold political pronouncements, it is nevertheless the case that most of them were welcome in the courts of kings and tyrants. It is instructive to compare the careers of the great geometers with those of the atomists, followers of Democritus (ca. 460–370 B.C.E.) and Epicurus (341–270 B.C.E.). The geometers, who practiced the science of universal order, were showered with favors by the great and powerful; the atomists, who believed everything in the world is a chance accumulation of randomly flowing atoms, were viewed with distrust and condemned as dangerous subversives.[9]

Still, on the whole it must be acknowledged that the impact of geometry on social and political life and thought in the ancient world was very limited. To some extent this is because mathematicians, then as now, tend to stay clear of the hurly-burly of politics, much preferring the solitude of their study and the exclusive company of fellow professionals. But the deeper reason is that even the admirers of the geometrical world considered it, for the most part, entirely separate from our familiar material reality. Consequently there was no reason to expect that social and political arrangements should correspond to the strict order of Euclid's dreamworld.

This view, that our actual world has little to nothing to do with the ideal geometrical one, has much to commend it. Our daily experience, after all, teaches us that our surroundings are diverse, irregular, unpredictable, and often irrational—the polar opposite of Euclid's geometrical universe. And if daily experience was not enough, there is the authority of the greatest philosophers of the ancient world to back it up. Aristotle (384–322 B.C.E.) was thoroughly skeptical of geometry's applicability to the physical world. Only the observation of particular objects in our messy irregular surroundings, he taught, could lead to knowledge about the world, and there is no reason to think that general laws so obtained would conform to geometry. Plato, for his part, had praised geometry not because it corresponded to our experience of the world, but for exactly the opposite reason: because it led us away from transitory incoherent matter and toward the perfect universe of the forms. Accordingly, geometrical order belongs not to our world at all but to the eternal forms. In our own corrupt world, which is but a shadow of the true one, perfect truth, order, and hierarchy will forever remain beyond reach.

And so while the *Elements*, and in particular the first six books dealing with plane geometry, were quickly recognized as a masterpiece, their impact beyond the scholarly world remained muted. The question of political order was a legitimate topic of philosophical discussion (though interest in political philosophy faded with the decline of the Greek city-states in the fourth century B.C.E.). It was certainly the domain of tradition, of religion, and, as is forever the case, of unapologetic power struggles. But it was not the domain of geometry, which belonged to a different realm, more beautiful and exalted than the one we inhabit. Consequently, the proofs and truths of geometry, impressive though they undoubtedly were, told us nothing about how to arrange our lives in the here and now.

The widespread conviction that geometry has little to do with the physical or human world survived the decline of the ancient world and the fall of the Western Roman Empire in the fifth century C.E. In the centuries that followed, as population levels crashed, cities were emptied of their inhabitants, and literacy rates plummeted, it was the Christian Church that became the sole guardian of culture and learning in Western Europe. This was more by default than by design, as Christian clerics were far from being fans of the great philosophers and scientists of antiquity. For centuries the Church had denounced their teachings, viewing them as dangerous rivals, better suppressed than preserved. But with literate culture itself on the verge of disappearing, the Church was simply the only institution with any interest in preserving the masterworks of Greek and Roman civilization, and the resources to devote to the task. And so while many churchmen cared nothing for philosophy and science, some nevertheless strove to preserve parts of the ancient tradition, and to adapt it to their purposes.

Geometry did not rank high in the Church's priorities. Euclid was known mostly through the works of Boethius (480–524 C.E.), who listed the theorems without proofs, whereas more advanced treatises, such as those of Archimedes and Apollonius, were completely lost for centuries. But Plato fared better: his view that our world is but a pale shadow of the perfect world of the forms resonated with Christian doctrine, and particularly with the teachings of the most influential of the Church Fathers, St. Augustine of Hippo (354–430 C.E.). Augustine believed that we humans are fallen beings who live in a corrupt and fallen world, and that our only hope for salvation was to transition to God's heavenly kingdom. Augustine's kingdom of heaven differed from Plato's world of the forms in that it could be entered only through God's grace, not by human reason. But when it came to the physical, material world, Augustine fully shared Plato's dim view: it was but a pale shadow of a higher and nobler world, be it the realm of the forms or God's heaven. And so, inasmuch as interest in geometry existed at all in the Dark Ages that followed the fall of Rome, it was understood within this Platonic framework: the truths of geometry may be very fine indeed; but our material world, the one we encounter and live in each day, is decidedly un-geometrical.

So things remained for the better part of a millennium. Even as

Western Europe emerged from the Dark Ages, as urban life reawakened in the eleventh century, and intellectual life revived in the twelfth; even as most of the geometrical tracts we know today, after being preserved for centuries in Arabic translation, were recovered by European scholars in the twelfth and thirteenth centuries, and Euclid was incorporated into the curriculum of the universities; even as some European algebraists, such as Leonardo of Pisa (also known as Fibonacci) and Jordanus of Nemore, produced original work of their own, attitudes toward geometry remained largely unchanged. This, on the whole, should not surprise us. For throughout this period that we know as the Middle Ages, the Church remained the heart and center of all European intellectual life; and the Church, arbitrator of all learning, remained decidedly unimpressed with geometry.

Church scholars, to be sure, did not deny that geometrical proofs established rational truths. Aristotle, after all, acknowledged as much, and Aristotle was the greatest intellectual authority of the Middle Ages. But where did that leave their flock? For them, geometry could not offer an account of the material or human worlds, fallen and corrupt as they are; nor could it offer a glimpse of heaven, or eternal truths, as it did for Plato, for those, according to the Church, depended entirely on divine revelation and God's grace, inscrutable to human understanding. And so, residing neither in the heavens nor on Earth, geometry in the Middle Ages was left suspended in ambiguity. For nearly a thousand years, from the fall of Rome to the year 1400, geometry was a perfectly rational universe that belonged absolutely nowhere.

The Geometrical Universe

Then, in a historical blink of an eye, everything changed for geometry. In the 1400s the Florentine polymath Leon Battista Alberti (1404–1472) claimed that geometrical order pervaded our world: the key to correct painting and architecture, he argued, was to reveal this hidden structure.[10] In the 1500s the Polish-German astronomer Nicolaus Copernicus extended geometry to the heavens as well: his Sun-centered system, he argued, was the true one, because it was geometrically superior to the traditional Earth-centered scheme.[11] And in the 1600s the Florentine scientist Galileo Galilei gave the most famous, and perhaps most beautiful, expression to the geometrical vision of the world. "This grand book, the universe," he wrote, "is written in the language of mathematics, and its characters are triangles, circles, and other geometrical figures, without which it is humanly impossible to understand a single word of it."[12] Galileo devoted his life to proving this, and paid the price when Church authorities objected to his placing geometrical arguments above scriptural authority.

It was a radical, even revolutionary, change. For a full millennium the Church had taught that our world was fallen, corrupt, and transient, and was therefore entirely un-geometrical. For another eight centuries before that the followers of Plato, who were geometry's most devoted advocates, had taught that geometry described only the exalted world of the forms, not its pale shadow, the physical world. And yet for all that, in the face of both religious and philosophical authority and a tradition of nearly two millennia, geometry was brought down to Earth, and suddenly infused everything around us. From the time of Alberti and Galileo onward, geometry seemed to be everywhere: it structured Earth and the heavens; it structured matter and motion; and as this book will show, it also structured society and the state. Geometry infused the natural world and the human one. It was the deepest order of the universe, the law that governs all.

The rapidly spreading conviction in the sixteenth and seventeenth centuries that our world was deeply geometrical has often been linked to the rise of modern science. There is good reason for this, for geometry was at the heart of some of the most spectacular accomplishments of the Scientific Revolution. Copernicus, who insisted that geometry described the physical heavens, was followed by Johannes Kepler (1571–1630), who discovered the true elliptical orbits of the planets and determined the geometrical laws that govern their motions. And it was Isaac Newton (1642–1727) who discovered the deep mathematical mechanism of planetary motion and presented it in strict Euclidean format in the *Principia Mathematica* of 1687. Galileo meanwhile investigated the geometrical structure of matter and the principles of motion, leading to his discovery of the law of falling bodies and ultimately to Newton's laws of motion.

These are among the most famous instances in what was in truth a geometrical tidal wave that carried all before it and set the scientific agenda for scholars throughout Europe. Nor did the impact wane in later times. Even today, scientists who seek out the mathematical laws that govern minute subatomic particles, or the incomprehensibly vast expanding universe, are still following the program set down by Galileo four centuries ago, when he declared that the world is written in the language of geometry.

But it was not only the natural world that was infused with geometry: it was the human world as well. For geometry presented a universe that was perfectly rational, impeccably ordered, and strictly hierarchical, one in which everything is forever just as it must be. As long as this universe was kept separate from our own, no lessons for the ordering of society could be drawn from that impeccable reality. But when our world became geometrical through and through, it was not just physical matter that was transformed but human creations as well: society, political institutions, and the state. This was, it seems, unavoidable. If geometry was the deepest order of the universe, it followed that the strongest, most harmonious, and most enduring social arrangements, the ones that were true and proper, were those that adhered to the geometrical model.

This book tells the story of the geometrical universe and how it shaped the social and political landscape, first in Europe, then far beyond. We begin on the streets of Florence in the early 1400s, where geometry was first brought down from its Platonic heavens. The discovery of the laws of linear perspective by Florentine artists and scholars revealed the deep geometrical architecture of space itself and transformed people's conception of their place within it. From there we move northward, to the Renaissance courts of the kings of France, who were the first of all the princes of Europe to recognize the social and political power latent in geometry. Hesitantly at first, but with increasing confidence as reign followed reign, the French monarchs modeled their kingdom on the orderly hierarchies of geometry. Their palaces and gardens became geometrical utopias, perfect worlds in which each object and each person occupy their God-given place in a strict hierarchy that led, inescapably, to the king himself.

The brilliant gardens of Versailles were the culmination of French royal geometries but also the starting point for the global spread of the geometrical ideal. Before long the Euclidean patterns of Versailles were carved into garden paths and city streets from Paris and Vienna to New Delhi and Manila. Wherever European powers established colonies and founded cities, there one would find triangles, circles, and ovals, the emblems of the Euclidean universe. Even the greatest republic in the world built its capital as the greatest of all geometrical cities. In its rigorous design, Washington, D.C., encodes a careful balance of power between the president and Congress, federal authority and the states, a strict republican order kept in place by the unchallengeable power of geometry.

The Seeds of Modernity

In the centuries since it was brought down to Earth, geometry has been adopted, utilized, and exploited by almost every conceivable art, profession, and political dogma. It was enlisted in support of French absolutism, American republicanism, and British imperialism, to name but a few of the geometrical political creeds. It has served science, but also philosophy, as treatises as diverse as Descartes's *Principles of Philosophy* (1644), Hobbes's *Leviathan* (1651), and Spinoza's *Ethics* (1677) modeled themselves on Euclid's *Elements*. To these one could add the ancient art of painting and the new one of ballet, the crafts of architecture and gardening, and more. There was, in fact, no limit to geometry's reach. As Bernard Le Bovier de Fontenelle (1657–1757), perpetual secretary of the Paris Academy of Sciences, put it in 1699, any "work on ethics, politics, criticism, and, perhaps, even rhetoric, will be better . . . if done by a geometer."[13]

The reason is clear. Geometry, for Fontenelle and his contemporaries, was not just a mathematical field with an impressive pedigree, but something far grander. It stood for the promise that the world in all its richness and variety, both natural and man-made, is, at its core, perfectly rational and perfectly ordered; that beneath the façade of chaos and irregularity that we see around us there is a deep unchanging order in which all things have their unique and incontestable place. So when philosophers claimed geometrical authority for their doctrine, painters insisted on the geometrical foundations of their art, and dance masters declared the geometrical underpinnings of their moves and positions, they were not simply providing an elegant and stylish organization for their disciplines. They were, rather, claiming that their creations—whether philosophies, paintings, dances, or anything else—were expressions of the deepest order of the universe.

Exactly the same was true for the political claims made in the name of geometry. When the kings of France created vast geometrical wonderlands on their royal estates, when George Washington insisted that the federal capital be built as a grand geometrical city, when the viceroy of India selected a geometrical design for the new capital of the British Raj, they were not just showing an aesthetic preference for straight lines, perfect circles, and precise angles. They were claiming that their political order—royalist, republican, or colonial—was an integral part of the deepest order of the universe. As such, it could never be overturned. For who, whether prince, peasant, or foreign potentate, could contest the rights of a king, overcome a republic, or topple an empire whose rule was founded on geometrical proof?

In the early 1400s geometry came down to Earth, bringing with it the promise of a rational and irrevocable universal order that reaches to all corners of creation. It made not only modern physical science possible, but also the modern state in all its variations—from kingdoms to republics to empires. Geometry made the modern world possible. This book tells its story.

PART I

≡

How the World Became Geometrical

1.

THE MIRROR IMAGE

The Gates of the Baptistery

By the early 1400s the Baptistery of St. John at the heart of Florence was already ancient, its origins lost in the mists of time. According to Giovanni Villani, the city's medieval chronicler, the structure was originally a Roman temple dedicated to Mars, the god of war. This seemed entirely appropriate to the proud Florentines, who believed their city had been founded by the rugged veterans of Julius Caesar's legions. Modern archaeology has revealed that this was almost certainly not the case, and that the Church of St. John (or San Giovanni, as it was known) was built from the ground up as a Christian house of worship. Beyond this, however, experts are divided even to this day: Was the striking octagonal edifice built in the late fifth century, when the Church sought to extend its reign as the Roman Empire crumbled around it? Was it a monument to the conversion of the Lombards, the Germanic tribe that ruled over much of Italy in the seventh century? Or was it an expression of burgeoning civic pride when, in the eleventh century, the city of Florence emerged from centuries of obscurity to become a bustling hub of commerce and culture?[1]

We do not know, as the thick walls of the baptistery hold tight to their secrets. For a thousand years the walls bore witness as each and every Florentine newborn passed through their gates. In a soothing rhythm of Christian life, men and women, rich and poor, commoners and aristocrats of ancient lineage, all were baptized in the shadow of those walls and joined together in a community of the faithful. Other events were far less tranquil, as, generation after generation, the walls stood silently by while the violent civic life of an Italian city-state raged around them. Beginning around the dawn of the second millennium C.E., as Florence rose in power and wealth, life in the city swung wildly between years of peace and prosperity and periods of brutal strife and civil war. The ancient families of the countryside, who ruled Florence during the Dark Ages, battled for power with the wealthy merchants and bankers who increasingly came to dominate the city's economic life. Rival clans of magnates turned their city dwellings into fortified towers and fought pitched battles in the streets, leading to the victory of some and the banishment of others. Guelphs, or champions of the Papacy, fought Ghibelines, who defended the rights of the German emperor, until the Ghibelines were finally driven out, never to return. And members of the working class, full-blooded Florentines who labored in the service of their wealthy brethren, rose up repeatedly to assert their political and economic rights. Wise to the danger, the squabbling clans of the city's elite would put aside differences long enough to crush the popular uprisings and ensure that the "republic" would continue to be ruled by those rich in money or land.

But on one summery day, in about the year 1413, the ancient walls of the baptistery were treated to a scene unlike any they had witnessed before.[2] It began unremarkably, as a short man in his mid-thirties, with a balding head and aquiline nose, marched briskly through the chill morning air and headed to the monumental doorway of the Duomo, across the piazza from the baptistery.

Known officially as the Cattedrale di Santa Maria del Fiore (Cathedral of St. Mary of the Flower), the Duomo was the pride of Florence—its main church and one of the largest and grandest in all of Christendom. Compared to the ancient and myth-shrouded baptistery, the Duomo was then practically new, a recent addition to the Florentine skyline and as yet an incomplete one. Nearly 120 years after its first stone was laid down, in 1296, the great cathedral was still missing its distinctive giant dome, so familiar to visitors today. Although it had been envisioned by the cathedral's architects as a key element in their design, no dome on such a scale had ever been built, and its construction had so far exceeded the capabilities of the best master craftsmen of the day.

In due time the man walking past the baptistery that morning would change all that, for he was Filippo Brunelleschi (1377–1446), who would one day make his name as the designer and builder of the great dome. But on that day in 1413 Brunelleschi had other matters on his mind: he walked directly under the great archway of the Duomo's main doorway as if he were about to enter, then abruptly turned around to face the baptistery. In his hand he held a modest painting, about one foot square, and a mirror of roughly the same size.

Anyone standing close to Brunelleschi that day in the shadow of the archway would have noted with surprise that the painting was not, as one might expect, a religious scene of the kind favored by the artists of the period. It was, rather, a simple depiction of the precise view from the spot where Brunelleschi was standing: the Baptistery of St. John as seen from the entrance to the Duomo, except that in the place where the sky would normally be the painting was coated with burnished silver, reflecting anything passing before it. Even more surprisingly, near the center of the painting, at the precise point depicting the baptistery's wall opposite the cathedral's doorway, there was a small hole. Brunelleschi raised the painting to his face and peered through the hole at the octagonal walls of the baptistery across the piazza. As puzzled onlookers watched, he raised the mirror and placed it in front of the painting, so that all he could see through the hole was the painting's own reflection. What, they must have wondered, was he doing?

While Brunelleschi in 1413 had yet to acquire the towering stature he would reach in later years, he was, nonetheless, already a well-known figure in Florence. A master goldsmith by trade, young Filippo got his first chance to make his mark in 1401, when at twenty-four he was a leading contender in a competition to design a set of giant bronze doors for the baptistery.[3] Brunelleschi's entry, depicting the sacrifice of Isaac and his rescue by God's angel, was fiercely expressive and impressed both the judges and the Florentine public. But he nevertheless lost the competition to the even younger craftsman Lorenzo Ghiberti (1378–1455), whose design was elegant and refined rather than dramatic.

The contrast between the two rivals extended beyond their artistic sensibilities. Ghiberti was not only a brilliant craftsman but also a sociable and amiable man with a penchant for quiet diplomacy. Throughout the competition he reached out to fellow artisans, consulted them, and incorporated their suggestions in his design. Brunelleschi, in contrast, had already earned a reputation as irascible, arrogant, and suspicious, a difficult man jealous of his methods and his credit. "To disclose too much of one's inventions and achievements is one and the same thing as to give up the fruit of one's ingenuity," he told the engineer Mariano di Jacopo Taccola years later, and there is no doubt he practiced what he preached.[4] He spent the year allotted for the competition working alone, in secrecy, never letting anyone but his closest companions see what he was doing. As a result, when the time came to pick the winner, Ghiberti had many friends among the judges and in the community at large, whereas the taciturn Brunelleschi was just as he liked it—alone.[5]

What happened next is much in dispute. Ghiberti, years later, claimed that he won the competition outright: "To me was conceded the palm of victory, by all the experts and by all those who had competed with me. Universally I was conceded the glory without exception." Yet exception there surely was: according to Brunelleschi's contemporary and biographer Antonio Manetti (1423–1497), when the time came to choose a winner the judges were already well familiar with Ghiberti's panels, and could not believe any of his competitors could do better. Once they saw Brunelleschi's striking design they realized their mistake, but felt unable to go back on what they had well-nigh announced—that Ghiberti was the clear winner. They settled on a compromise, declaring both men winners and commissioning them to work together. Ghiberti agreed; Brunelleschi, characteristically, refused, and walked away from the project, leaving the design and casting of the baptistery doors in his rival's hands.[6]

The Spell of the Ancients

Devastated by his loss in a competition he thought he deserved to win, Brunelleschi did not linger long in the shadow of his triumphant rival. Instead, he traveled to Rome, where he spent much of the next fifteen years far from the jealousies and rivalries of his native city. And if disappointment was enough to keep him away from the city on the Arno, it was something else that drew him inexorably to Rome: an obsession with for the ancient classical civilizations, which Brunelleschi shared with many of his most distinguished contemporaries.

This passion for ancient Greece and Rome had spawned a movement known as humanism, which had begun in the previous century but was now sweeping through Italy and reshaping the intellectual landscape.[7] For the medieval schoolmen, ensconced in the famous European universities, the humanists had little but contempt. For all their learning, the schoolmen, in the humanists' opinion, relied almost exclusively on a single ancient source, the writings of Aristotle, which they had by secondary (and, according to the humanists, corrupt) translation from Arabic. Even worse, the schoolmen's very language, medieval Latin, was but a pale shadow of the rich and flowery language of Cicero and Livy. Little wonder that the schoolmen were obsessed with abstruse Aristotelian commentary and pointless theological debate. On the questions that the humanists believed truly mattered—how to live a good, moral, and worthy life—the medieval schoolmen were silent. There was nothing for the humanists but to make a clean break with their medieval forefathers and draw directly from the ancients.[8]

In their quest to recover ancient learning, the humanists studied philosophers such as Plato and Seneca, scientists such as Archimedes and Ptolemy, historians such as Polybius and Tacitus, and poets including Virgil and Ovid. But of all the ancient authors none was more greatly admired than the Roman statesman Marcus Tullius Cicero (106–43 B.C.E.), who embodied the humanist ideal to perfection. Cicero was not only the greatest of Roman orators, and a moral philosopher of note, but also a man who put his teachings to the test. A lifelong participant in Rome's cutthroat politics, he saved the state from a dangerous conspiracy while serving as consul and was hailed by the Senate as Pater Patriae ("Father of the Country"). One could ask for no finer model of linguistic purity, literary prowess, and civic engagement.

FOR THE MOST PART, HUMANISM was a literary and philosophical movement focused on books and texts. Itinerant scholars such as Poggio Bracciolini traveled far and wide in an effort to locate lost ancient works that might be hidden away in the monastic libraries of Europe. Seeking out the most authentic versions, they worked hard to restore the texts to their pristine glory in their original tongues, mostly Greek and classical Latin, but also Hebrew, Arabic, and other languages. These newly recovered texts would then circulate among the humanists, replacing existing (and allegedly corrupt) medieval versions if such existed, or adding new ancient sources if they did not.

Brunelleschi, however, was not a literary scholar. A brilliant artist, architect, and engineer, he was not a man to spend his days poring over ancient manuscripts. For his humanist friends the passion for the ancients meant recovering the writings of Plato, Cicero, and Lucretius and returning them to their original brilliance. For the practical-minded Brunelleschi it meant something else: studying the physical traces of the ancients—the buildings, aqueducts, roads, and sculptures that they left behind. The humanists traveled to far-flung corners of Europe in pursuit of their original texts. Brunelleschi, in contrast, headed straight to the capital of the ancient world and the home of its greatest monuments: Rome.[9]

At the turn of the fifteenth century, Rome had sunk to one of the lowest points in its long and illustrious history. Back in the first century C.E., as the capital of a great empire, the city had boasted a population approaching two million, making it by far the largest city in the Mediterranean world, and probably anywhere. But four centuries later, with the empire in steep decline, the population had dwindled to less than half that number. The barbarian invasions that followed, accompanied by the collapse of commerce throughout Western Europe in the early Middle Ages, saw the city's population crash to a tiny fraction of its ancient heights. Even the revival of urban life in the eleventh and twelfth centuries, which saw ancient towns come to life and new ones spring up throughout Italy, did little to improve the fortunes of Rome. While cities such as Florence and Venice grew into bustling centers of commerce and manufacture, the ancient imperial capital fell further and further behind. In time the popes of the Renaissance would rebuild and repopulate the city and make it a worthy capital of their spiritual empire. But in 1403, when Brunelleschi arrived, Rome was a poor and disease-ridden town of perhaps 30,000 souls, many of whom lived like vagabonds among the monumental ruins.

The young Florentine, however, had no interest in the sad, dilapidated Rome of his day. All he saw was the city's past greatness: the remains of the great temples in the Forum; the seemingly indestructible roads, many still in use; the ruins of the massive aqueducts that had supplied water to the city of millions. Among the great structures still standing he surely would have noted the circular Colosseum, built by the emperors Vespasian and Titus in the first century C.E., and the monumental public baths built by the emperor Diocletian around 300 C.E., renowned for their high, vaulted ceilings. Most striking of all was the Pantheon, with its massive dome, whose span and height of 142 feet was still the largest in the world.

Brunelleschi, to be sure, did not come simply to admire the faded glory of the Eternal City: he was there to learn everything he could about the structures of ancient Rome, from simple abodes to great monuments, and to determine how they were built. As his biographer Vasari told it more than a century later, "his studies were so intense that his mind was capable of imagining how Rome once appeared even before the city fell into ruins."[10] If the humanists' dream was to recover the intellectual world of Cicero, Seneca, and Tacitus, Brunelleschi's was to recover their physical world—the streets they walked, the houses they inhabited, the temples in which they worshipped.

Rather than work alone, as was his habit, Brunelleschi this time was accompanied by a friend, the sculptor Donatello. Together the two Florentines set to work surveying every structure they could find: they "made rough drawings of almost all the buildings in Rome . . . with measurements of the widths and heights as far as they were able to ascertain," Manetti tells us. Since the structures were in ruins, and the original street level was buried deep in the ground, this was easier said than done. It often proved necessary to dig up the structures' foundations in order to determine their original shape and true height. When the job was too great for the two of them, they hired laborers, and when measuring heights and distances directly was impractical, they relied on geometrical measurement techniques, using surveying rods and mirrors. To record the measurements "they drew the elevations on strips of parchment graphs with numbers and symbols that Filippo alone understood." And so they surveyed house after house, and ruin after ruin, over the entire ancient city.[11]

Day and night the two Florentines labored on their colossal undertaking, but no one in the city understood what they were up to with all this digging and measuring. Lacking a better explanation, the Romans assumed the two were looking for gold or marketable goods, and so they became known in the city as "the treasure hunters." The Romans may not have been the only ones confused, since according to Manetti even Donatello was partly in the dark about Brunelleschi's true purpose. Secretiveness was to Brunelleschi second nature, and, as far as was possible, he kept his ideas from even his closest companion.

What, then, was Brunelleschi hoping to accomplish with his exhaustive survey of Roman antiquities? One goal, undoubtedly, was a professional one: Brunelleschi wanted to learn the construction methods of the ancient Romans so that he could make use of them in his own projects. For example, according to Manetti, "he considered the methods of centering the vaults and other systems of support, how they could be dispensed with and what method had to be used, and when."[12] Perhaps he was already devising the machines that he would use, years later, to build the great dome of Florence's cathedral.

But there was something more to Brunelleschi's "treasure hunt" than the acquisition of technical knowledge. Brunelleschi believed that the ancient architects designed their structures in accordance with perfect mathematical harmonies, which were responsible not only for the soundness of the buildings, but also for their beauty. His goal, according to Manetti, was "to rediscover the fine and highly skilled method of building and the harmonious proportions of the ancients."[13] Some of these harmonies were fairly straightforward: the interior of the dome of the Pantheon was an almost perfect half-sphere, and the classic columns used in all public buildings were constructed according to strict mathematical proportions. Ionic columns were nine times as high as the diameter of their base, Corinthian columns ten times their base, and so on.

Other symmetries and harmonies, however, could not be so easily discerned. Brunelleschi believed that the buildings of Rome harbored many more secrets, which could be uncovered only through painstaking systematic work. By surveying, measuring, and recording the heights, lengths, shapes, and distances of every ancient structure in Rome, he was hoping to uncover those hidden mathematical harmonies that guided the ancient builders of the Eternal City and had since been lost. In the hills of Rome others might see a haphazard and disorderly jumble of ruins in various stages of disrepair. Brunelleschi saw the outlines of a perfect, harmonious order that pervaded every structure—and the entire city.

The View from the Duomo

Even while still spending most of his days as a Roman "treasure hunter," Brunelleschi made regular visits to his native city, where he remained a familiar figure. And it was one of those visits, in or around 1413, on one morning when we find him walking briskly past the baptistery, site of painful memories, and toward the doors of the Duomo. He may well have been contemplating the great cathedral's missing cupola, while nurturing a secret ambition that he would be the man to build it.

Yet the objects he was carrying that morning were not the tools of a goldsmith and caster of bronze, as one would expect of the master craftsman who had lost the competition for the baptistery's doors by the slimmest of margins. Nor were they architectural designs or engineering plans, as later generations, who know him as the builder of "Il Cupolone," might expect. A mirror and a painting with burnished silver and a hole in the middle are all that Brunelleschi held in his hand. In what must have been no more than a few hours, and using these simple objects, he forever transformed the way people perceive and experience the space around them.[14]

Here's what Brunelleschi did: First, he held the back of the painting to his face, placing the hole in its center before his eye. Manetti, who held the painting in his hands decades later, reported that the hole was "as tiny as a lentil bean on the painted side" and that it "widened conically" on the back side, reaching "the circumference of a ducat."[15] Brunelleschi pressed his eye to the cone and looked through the hole toward the baptistery. With his other hand he held up the mirror in front of the painting so that, instead of the actual octagonal structure, he saw its image in the painting reflected back at him. There was the baptistery and the buildings around it as he had painted them. There was the blue-gray sky, the drifting clouds, and passing birds captured in the burnished silver on the painting and now reflected from the mirror and back at Brunelleschi's eye. From the peephole at the center of the painting, Brunelleschi was looking straight at the painting's reflection in the mirror. The artificial image was almost indistinguishable from the view of the actual baptistery as seen from the same spot.[16]

It was surely one of the most baffling scenes ever to take place before the ancient walls of the baptistery. From a distance it would have appeared that Brunelleschi was inexplicably covering his own face with the back of a painting; from close by, when the hole in the painting was revealed, it seemed that he was going to enormous lengths to see what was freely available to the naked eye—the octagonal outlines of the baptistery in the morning light. Why the painting, the peephole, the mirror, if the only purpose was to reproduce the exact scene that could be viewed without them? What, one might understandably wonder, was Brunelleschi up to?

The answer lies precisely in the bewildering identity of the natural view from the cathedral's doorway and the artificial view of the painting and mirror that Brunelleschi worked hard to achieve. The whole purpose of the experiment, in fact, was to demonstrate that he had mastered the secret of the true representation of nature, one that differs not at all from the real thing. The more similar the mirror image of the baptistery—seen through the hole in the painting—is to the actual view from the Duomo's doorway, the better. The best indication of the experiment's success was if the mirror was suddenly yanked away, the view through the hole in the painting remained practically unchanged. Brunelleschi had succeeded in capturing a three-dimensional view of an object and then reproducing it from scratch on a flat canvas. Nothing like it had been accomplished before, but his contemporaries soon had a name for it: they called it *perspective*.[17]

That Brunelleschi was widely considered the founder of perspective we know from his contemporaries: not only his biographers, Manetti and Vasari, but also the humanist Domenico da Prato, who wrote admiringly about "the perspective expert, ingenious man, Filippo di Ser Brunellesco, remarkable for skill and fame."[18] And Manetti's friend Cristoforo Landino reports that "the architect Brunelleschi was also very good at painting and sculpture; in particular, he understood perspective well, and some say he was either the inventor or rediscoverer of it."[19] Sadly, however, Brunelleschi's painting of the baptistery is lost to us. Manetti, writing decades later, tells us that he had held it in his hands and "seen it many times," and it was listed among the effects of Lorenzo the Magnificent, Medici ruler of Florence, after his death in 1492. But that is the last we hear of it, and as a result we cannot know with certainty what exactly Brunelleschi did in the painting to earn him his reputation as the founder of perspective. Yet his use of a mirror in his experiment provides an important clue.[20]

Introduced to Europe in the thirteenth century, mirrors were still rare and expensive items in Quattrocento Italy, the objects of widespread fascination. A mirror, after all, is a flat, two-dimensional surface, just like a wall or a painting. And yet when we look at a mirror it is almost impossible to maintain our awareness that the mirror is flat, because the perception of depth practically overwhelms us. We see our own reflection not on the surface of the mirror but behind it, just as we see all the objects around us deep in the mirror's illusory space. Furthermore, all objects appear at the same distance "behind" the mirror as they are before the mirror's surface.[21]

Today we take mirrors and their images so much for granted that we hardly think them worthy of comment. Yet there is magic here: without any human intervention, mirrors produce a perfect three-dimensional image on a flat surface, one that is practically indistinguishable from the original. How do they do it? And what does an artist, using all the skill and human artifice at his disposal, need to do to reproduce the effect? This seems to have been the question on Brunelleschi's mind as he was designing his experiment. If he could reproduce a mirrorlike sense of depth in his painting of the baptistery, then the image the flat painting reflected back at the observer would appear not flat at all but three-dimensional. For an observer, it would be as if they were looking directly at the baptistery.

To experience firsthand how a mirror produces its depth effect, try a simple experiment: stand directly in front of a mirror at your home alongside at least two parallel objects perpendicular to the surface of the mirror. The traditional parallels are railway tracks or the sides of a road, but since you are unlikely to find these in your house, then towel or clothes racks, or the outlines of furniture or door frames, will do just as well. Using a digital camera, take a photograph of the image in the mirror, including yourself, the camera, and the various parallel objects. Then inspect the resulting picture, and with a ruler extend the lines of the parallel racks (or furniture, or door frames) until they intersect. You will find that all the lines meet exactly—but exactly!—at the point occupied in the picture by the lens of the camera. If we move the camera to a different location—higher or lower, left or right—and take the picture from there, the result will be exactly the same: the intersecting lines will follow the camera lens wherever it goes. In a mirror, it is fair to conclude, all parallel lines perpendicular to its surface meet at a single point—the one occupied by the observing lens.

Back in 1413 Brunelleschi did not have the benefit of cameras, digital or otherwise, but had to observe the image on a mirror directly. This is not easy to do, because looking at a mirror we are so powerfully drawn into its three-dimensional world that it is next to impossible to conceive of it as a flat surface. Yet Brunelleschi, making good use of his experience using mirrors in his Roman survey and translating structures into two-dimensional diagrams, apparently managed to do so: in a mirror, he discovered, all parallel lines meet at a single point. In his case, of course, the point was occupied not by an artificial camera lens but by the observer's own human eye.

It was a remarkable discovery, and Brunelleschi clearly believed that it held the key to replicating the "mirror effect" in a work of art. If a painter, through his skill, could artificially reproduce what a mirror did naturally, then surely the surface of the painting would appear just as transparent and three-dimensional as the surface of a mirror. There is every reason to believe that that is precisely the effect Brunelleschi was aiming for when he painted the image of St. John's baptistery as seen from the portico of the Duomo. In his painting the parallel sides of the portico where he was standing, the parallel rows of cobblestones on the pavement between the Duomo and baptistery, the parallel lines of the roofs at the edge of the piazza on both sides of the baptistery, all converged on a single point in the painting. It is what is known today as the *vanishing point.*

In a mirror the vanishing point is located at the observer's eye as seen in the glass. In a painting, which usually does not include an image of the observer, the point is defined by the place where the observer "would have been" had the canvas been a mirror instead: it is an eye-level point on the canvas, directly across from where the observer should be standing before it. In the case of the baptistery painting, that is in the middle of the structure's eastern doors, the ones facing the cathedral.[22] By placing his peephole in this location, Brunelleschi was reproducing the precise effect he had observed in mirrors, where

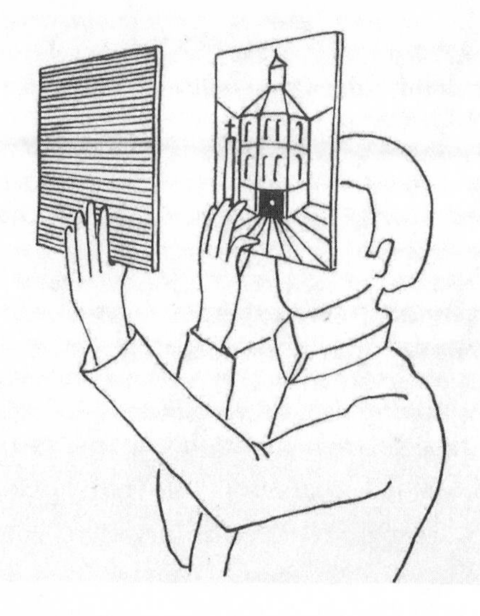

FIGURE 4: **Brunelleschi's experiment**

all parallel lines converged onto the observer's eye. When he looked through the hole at a mirror held before him, the flat surface of the painting disappeared and was replaced by a deep three-dimensional image of the baptistery and the piazza. Without the vanishing point, a painting is like a wall—an opaque barrier to vision, hiding what lies behind it and replacing it with jottings on a flat surface. But the vanishing point transforms a painting into a window, which does not block our view but rather opens up an alternate reality behind it and invites us in.

In Rome, Brunelleschi had spent years searching for the hidden geometrical principles that the ancients used to build their city. Years later he would make use of what he learned in his architectural work. The Church of San Lorenzo, for example, which he designed and began building simultaneously with the cathedral's great dome, was constructed almost entirely of standard square units, combined in different proportions of whole numbers. Most famously, it was geometrical harmony that inspired the aesthetics of Il Cupolone, just as the geometrical principles of engineering ensured that the construction was possible and the dome structurally sound.[23]

In the perspective experiment Brunelleschi was expanding the reach of geometry even further. Here was a geometrical secret that was not simply a matter of human artifice, like the monuments of the Eternal City or the great dome: it existed in nature itself, and manifested itself in the reflections of mirrors. Brunelleschi did not invent the principles of perspective but merely revealed a natural geometrical order that was already there. It wasn't simply that men could use geometry to create great and wondrous objects. Instead, Brunelleschi's experiment suggested, the world around us is already imbued with geometrical patterns that we need only reveal. Nature itself is geometrical.

If the implications of Brunelleschi's discovery had been perceived at the time, they could have shaken the foundations of his world. Ever since its invention two thousand years before, geometry had been praised for its logical rigor and admired as a model for attaining true knowledge. Plato, who lived in the fourth century B.C.E., considered it the ideal science and a model for men seeking a glimpse of the pure and rational world of the forms, which, he believed, were the only things that truly existed. When, a few decades after Plato's death, geometry was systematized and codified by Euclid of Alexandria, it became the embodiment of absolute, unshakable truth.

This, however, did not mean that the world itself was geometrical! Far from it. Plato thought the physical world was a corrupt version of the rational beauty of the realm of the forms, and was therefore entirely unsuitable for geometry. The Christian Church, which dominated the intellectual world of the Middle Ages, agreed. As punishment for original sin, men live in a fallen world of falsehood and confusion, a world that would never conform to the rational dictates of geometry. Finally, Aristotle, the greatest intellectual authority in Western Christendom, believed that knowledge of the world should be acquired from experience, and he was highly skeptical as to whether such knowledge would conform to the rules of geometry. In the Middle Ages, broadly speaking, geometry was often praised and broadly admired; but the world itself was decidedly un-geometrical.

Brunelleschi's discovery challenged all that: here was a geometrical secret hidden deep among nature's inscrutable mysteries. Nature, it turned out, could indeed follow the rules of geometry, and if this was the case for perspective, was it not plausible that other geometrical secrets are as yet hidden and awaiting their discoverer? Despite the teachings of Aristotle and the Church, Brunelleschi's discovery hinted that the natural world might indeed be perfectly geometrical. Underlying the boisterous chaos and variety that we see around us there may yet be a rational order, which humans can comprehend and even imitate. It is hard to say how Brunelleschi himself perceived his discovery. It may have been, to him, no more than a clever trick that artists might use to dazzle clients and spectators. But the potential implications of the discovery were immeasurably greater, and over time they would upend Europeans' entire understanding of their world.

In the immediate aftermath, few paid much attention to Filippo's strange trials in the Piazza del Duomo. Shortly after the trial at the baptistery, Brunelleschi produced another perspectival painting of the neighboring Piazza della Signoria, apparently using a more complex method that included oblique vanishing points.[24] But that seemed to be the end of the matter, and apart from Domenico da Prato's reference to him as the "perspective expert," many years pass before we hear any more about it. It was only a decade later, in the 1420s, that paintings using Brunelleschi's method began adorning the churches of Florence.

The man most responsible for introducing and disseminating the principles of linear perspective was the painter Tommaso di Ser Giovanni di Simone, universally known as Masaccio (1401–1428). A full generation Brunelleschi's junior, he met the great architect in the early 1420s, when Filippo was the most powerful artist-engineer in Florence and Masaccio was a young painter from the provinces trying to make his way in the great city. The two formed a bond and, according to Vasari, Brunelleschi "worked very hard for a long time to teach Masaccio many of the techniques of perspective and architecture." In 1423, likely at his mentor's urging, Masaccio traveled to Rome, just as Brunelleschi himself had done two decades before. It was in the few short years between his return to Florence and his death at the age of twenty-seven that he produced the paintings that would establish him as one of the greatest artists of the Renaissance.[25]

The Invention of "Inner Space"

In 1425 the Dominican friars of Florence commissioned a fresco from Masaccio for their church of Santa Maria Novella, where the work, known as *The Holy Trinity*, can still be seen today. The main themes of the mural are no different from those of many medieval church paintings and would have been familiar to good Christians everywhere. At the center of the painting is Christ on the cross, set up in a hall with a barrel-shaped vault supported by columns, and flanked on either side by the Virgin Mary and John the Baptist. Above the cross looms God the Father, looking fixedly at the viewer and supporting the cross with his hands. Below the hall, and seemingly below ground, a skeleton lies atop a stone sarcophagus. The message was easily understood by members of the congregation, steeped as they were in the teachings of the Church: Christ transcends the certainty of death, and through his suffering extends a promise of eternal life.

But it is not the theological lesson or the main figures that make the painting so strikingly different from any that had come before it. It is, rather, the hall aboveground and the sepulcher below, the very space that the figures inhabit. In creating this space Masaccio followed Brunelleschi's method so closely that the painting almost appears to be a formal exercise in linear perspective. Most strikingly to the viewer, the barrel vault extends directly from the surface "into" the painting and is divided into identical squares laid out in regular parallel rows. Following Brunelleschi's guidelines, the parallels formed by the squares all meet at a single point, below the base of the cross. And similarly (if less obviously), the parallel sides of the subterranean sarcophagus also meet at the same vanishing point. Just as in the baptistery experiment, Masaccio positioned the vanishing point at eye level, about five feet and nine inches above the floor. The churchgoer at Santa Maria Novella is in precisely the same position as Brunelleschi was when he looked across the Piazza del Duomo at a point in the center of the baptistery doors. In the *Trinity* Masaccio had reproduced the mirror effect his teacher had tried to replicate years before. As in a mirror, the space of the painting,

FIGURE 5: **Masaccio,** ***The Holy Trinity*** **(ca. 1425)**

of the vault and the sepulcher, becomes an extension of our own space and sucks us into its imaginary world.

Unlike any work of art before it, the *Trinity* makes explicit use of Brunelleschi's principles, thereby creating a three-dimensional space within the painting.[26] Yet, innovative as Masaccio's technique was, there is no denying that the overall effect is static and rigid. The figures, while clearly inhabiting precise positions within the painting's inner space, also seem as stationary and immovable as the columns around

FIGURE 6: **Masaccio, *The Tribute Money* (ca. 1427)**

them, and not at all like living, breathing human beings. It would be two more years before Masaccio managed to produce lively and dynamic paintings that nevertheless fully adhered to the geometrical principles of perspective.

Around 1427 Masaccio executed a cycle of frescoes for the Brancacci chapel in the church of Santa Maria del Carmine in Florence, devoted to the life of St. Peter. The most famous of these is *The Tribute Money*, depicting how Jesus sent Peter to pluck a fish from the water and present the coin he would find in its mouth to the tax collector. Here again Brunelleschi's principles are dutifully followed, but with a far lighter touch than in the *Trinity*. In place of the all-encompassing vault, *The Tribute Money* uses only a single structure on the right side of the composition to define the vanishing point. In place of the rows of squares that dominate the *Trinity*, here there are only a few short parallel lines of the balcony and the floor, just enough to produce the illusion of depth, but no more. In addition Masaccio seems to have mastered a principle that was implicit in Brunelleschi's method: that the heads of the scene's protagonists (and any objects at eye level), regardless of their apparent

FIGURE 7: **Lorenzo Monaco,** ***Adoration of the Magi*** **(1420–1422)**

distance, should all be at the same height in the painting—that of the vanishing point.[27]

The result is a painting that has long been recognized as one of the masterpieces of the early Renaissance. Instead of forcing a geometrical straitjacket on the composition, the rules of perspective do just enough to bring the disparate elements into an elegant and harmonious whole. Linear perspective here is not an imposition on the painting but a resource ingeniously used to hold the overall composition together. In the two years between 1425 and 1427 Masaccio had transformed himself from a student of linear perspective to its master.

It is instructive to compare *The Tribute Money* with a work by a different Florentine artist, Lorenzo Monaco: *Adoration of the Magi*, painted only a few years earlier. The composition of Monaco's painting is as striking as that of Masaccio's—the figures are as elegant, the emotions as powerful, and the message as clear. Both paintings foreground the central narrative and position the main figures alongside an architectural structure, and in both the background is taken up with a forbidding mountainscape. And yet the overall effect is strikingly different: the swarm of figures in the *Adoration* crowd the surface of the painting and the forbidding mountain in the back serves as a final backstop that excludes any further extension of the painting's inner space. In fact, the painting's space exists only inasmuch as it is occupied by people and objects, and even the small areas of sky are sealed off with impenetrable gold.

In contrast, Jesus and his disciples in *The Tribute Money* fully inhabit the interior of the painting, whose depth extends into the distance, along the water and into the sky. Space here exists in itself, regardless of whether the figures and structures inhabit it or not. It is an absolute space established by the universal laws of geometry, and it precedes anything that takes place within it, even the miracles of Christ and the apostles. Only a few short years separate Lorenzo Monaco's *Adoration of the Magi* from Masaccio's *The Tribute Money*, and yet the two paintings seem to exist in different worlds. In the universe of the *Adoration* things are perceived in an imprecise jumble, much as they are in daily life, their location determined by their positions relative to one another. In the world of *The Tribute Money* all objects are precisely located within a preexisting space that extends to infinity. Monaco's world was as it had been for his medieval predecessors—tactile, rich, and sensuous. But Masaccio's world was different from anything that came before him: it was geometrical through and through.

2.

THE GEOMETRICAL CODE

The Exile's Return

As far back as the eleventh century the Counts Alberti were the mightiest noblemen of the Tuscan countryside around Florence. Feudal magnates that they were, the Alberti viewed with suspicion the bustling city that had arisen amidst their lands, and they did their best to limit its growth and power. They did not succeed: despite their vast estates and numerous retainers they proved no match for the dynamic Florentine republic, which soon reduced their citadels, crushed their independence, and forced the humbled counts to take up residence within the city, where they could be kept under close watch. Tamed but not broken, the Alberti recovered from their humiliation and prospered in their new surroundings. By the 1300s they were once again one of the richest clans in

Tuscany, though now their wealth was based not on landholdings but on a vast commercial empire that stretched from Constantinople to London.[1]

All seemed well for the proud Alberti until the 1380s, when the clan's elder, Benedetto degli Alberti, became entangled in one of the family feuds that were the bane of city-state politics. His enemy, unfortunately, was Maso degli Albizzi, the leader of the ruling oligarchy and the most powerful man in Florence. Once again, the Alberti were outmatched: in 1387 first Benedetto and then the rest of the clan were banished from Florence and dispersed across the Italian peninsula. And that is why Leon Battista Alberti, though a Florentine through and through, was born in Genoa, where his father had set up his household and business. In due time the young Alberti would become a humanist, a scholar, an architect, and the original "Renaissance man." And it is likely not a coincidence that it was he, high aristocrat that he was, and not the humble if respectable Brunelleschi, who wrote the first systematic treatise on perspective. For it may have required a man of Alberti's exalted social standing and far-reaching connections to transform perspective from a practical trick used by painters to a universal science studied and taught by scholars throughout Europe.[2]

Battista (he added "Leon" later in life) was the illegitimate child of Lorenzo degli Alberti, who set up his household and business in the Genoese Republic following his banishment from Florence. Despite the murky circumstances of Battista's birth, Lorenzo took a liking to his bastard son and provided him with an excellent education, sending the boy first to Padua, where he read the classics, and then to the ancient University of Bologna, where he studied law. But when his father died young Battista fell on hard times: his relatives refused to acknowledge his claim to the family fortune, leaving him without resources or protection. Driven by necessity, he took up a position as a clerk in the pope's entourage, and from that day on spent his life moving between the most powerful courts in Italy. To the popes in Rome, the Este of Ferrara, the Gonzaga of Mantua, and Montefeltro of Urbino, he served as a confidante, a secretary, and a public intellectual whose beautiful Latin and Italian prose lent glamour and prestige to his patrons. In later years, as an acknowledged master architect, he was also commissioned to design and execute a large enough number of palazzos, chapels, churches, and city plans that a modern scholar has referred to him as the "master builder of the Italian Renaissance."[3] In 1428 the Albizzi at long last fell from power in Florence and the ban on the Alberti clan was lifted. It was then that Leon Battista could finally add another name to his brilliant list of patrons: the Medici of Florence.

By the time Alberti first set foot in the city on the Arno in 1434, he was no longer an anonymous clerk in the papal service, but rather a rising star among the humanists of his day.[4] In fact he had just begun writing *Della Famiglia* (*On the Family*), a work that would establish his reputation as a leading scholar throughout Italy.[5] If *Della Famiglia* was considered novel by contemporaries, it was not because of the dialogue format (which was traditional) or even because of Alberti's decision to write in the vernacular rather than in scholarly Latin. Fellow Florentines Dante, Petrarch, and Bocaccio, after all, had done so a century before. It was, rather, the topic itself—family life—that made the work so strikingly refreshing. Here was a work by a distinguished and learned scholar, and yet it was devoted to the most mundane issues: how to pick a wife, how to maintain family harmony, how to preserve the family fortune. Given Alberti's difficult personal history those topics were likely close to his heart, but they were not the kind of questions that merited discussion in learned treatises. A scholar of his stature might be expected to write about theology, philosophy, or even science. But family life? What could a scholar possibly say about a subject as lowly as that?

Quite a bit, according to Alberti. By making family life the subject of his book he was giving practical advice on questions that were undoubtedly on the minds of many people of his social class. But he was also doing more: he was elevating daily life, making it a topic worthy of learned discussion. Like other humanists of his day he despised the medieval schoolmen and their intricate treatises on topics that seemed to have no bearing on people's actual lives, and he admired the ancient philosophers and their concern with human conduct and happiness. Alberti's older contemporary Leonardo Bruni (1370–1444) used these concerns as a launching pad for a patriotic history of Florence, glorifying the spirit of its people and their love of liberty. Some decades later the desire to make philosophy relevant motivated Pico della Mirandola (1463–1494) to write his "Essay on the Dignity of Man." But for Alberti this was not enough: taking humanist sentiment to its logical conclusion, he asserted that even the choosing of a mate and practicalities of a family business were subjects worthy of scholarly attention. Philosophical discussions, he implied, should be recalled from the lofty clouds of abstraction and made to work in the real world. It would not be long before he applied the same reasoning to another field that was considered too abstract and far removed to be useful in the real world: geometry.

Alberti's Window

In 1434, when Alberti arrived in Florence, it was the first time he had ever laid eyes on the Duomo, the Baptistery of St. John, the Palazzo della Signoria, the Ponte Vecchio. And yet it was also unmistakably a homecoming, ending what he called "the long exile in which we Albertis have grown old."[6] Young Battista moved comfortably among the city's elite, and it did not take him long to recognize that Florence was in the grip of an unprecedented artistic ferment. With Brunelleschi's massive new dome gracing the cathedral and the city's skylines, Ghiberti's "gates of paradise" adorning the baptistery doors, Donatello producing freestanding naturalistic sculptures, and Masaccio's striking perspectival paintings adorning churches and chapels, Alberti had no doubt that he had stumbled upon a unique historical moment. He also knew that he wanted to be part of it.

And so, within a year of his return, Alberti published *Della Pittura* (*On Painting*), a short treatise on the art of painting. He dedicated it to his new friend and the man of the hour in Florence, Filippo Brunelleschi. Although the treatise is divided into three "books," it is the first of these that has made *Della Pittura* the best known and most influential of Alberti's many works. This is where Alberti lays out his geometrical theory of perspective and provides detailed instructions on how to systematically produce a three-dimensional effect in paintings. Humanist that he was, Alberti fervently believed that higher learning should not be the insular domain of scholars but should instead infuse and illuminate the real world. Just as in *Della Famiglia* he brought lofty philosophy to bear on humble family life, in *Della Pittura*, written in the same years, he forged pure, exalted geometry into a tool in the hands of painters.

Since Book I of *Della Pittura* is, as Alberti explains in the dedication, mathematical in nature, it takes its inspiration from the greatest mathematical textbook of all time—Euclid's *Elements*. Just like the *Elements*, it begins with the definitions of the simplest things—a point, a line, a surface—and proceeds from there. There is, however, a crucial difference: "Mathematicians," Alberti explains, "measure the shapes and forms of things in the mind alone and divorced entirely from matter." But "we . . . who wish to talk of things that are visible, will express ourselves in cruder terms."[7] He is, he explains, speaking not as a mathematician but as a painter. His purpose is not to construct a beautiful abstract mathematical structure in the manner of Euclid, but instead actual physical paintings depicting the real natural world.

Following the elementary definitions, Alberti moves on to a discussion of the nature of sight, based on traditional medieval sources. Carefully avoiding the intractable ancient question of whether the eyes emit their own visual rays or receive those emanating from objects, Alberti defines the visual pyramid, with its base the observed surface area and its apex at the eye. For example, looking out at a garden the actual area of the garden is the base of the pyramid, with the rays all converging on the eye, its apex. Now, a painting, Alberti continues, "is the intersection of a visual pyramid at a given distance."[8] This means that as the rays converge from the external surface toward the eye, they occupy a smaller and smaller area. A painting is a slice of this pyramid, in which the rays have already converged to a certain extent and now occupy a smaller area—the surface of the painting. It serves, Alberti notes, as an "open window" into an outside reality, smaller than the actual objects beyond it but nevertheless presenting them precisely and correctly.

Next, Alberti proves geometrically that proportionate sizes of the observed objects and the distances between them will be preserved in the slice of the visual pyramid that is the painting. Consequently, although the surface of the painting is smaller than the actual area it depicts, it will appear the same to the eye, just as the surface of a small window can correctly show the much larger world beyond it. For if "the sky, the stars, the seas, the mountains, and all living creatures, together with all other objects, were . . . reduced to half their size," Alberti explains, "everything that we see would in no respect appear to be diminished from what it is now."[9]

So far, Alberti has offered little that was new to a contemporaneous scholar of the science of vision. *Della Pittura* is, to be sure, unique in addressing itself to practicing painters, not cloistered scholars, and in treating them as a worthy audience. The treatise was also novel in its unquestioned assumption that the goal of painting is a true representation of the world, and that to accomplish this it must be based on strict geometrical proportions. But the actual theory of vision presented was, in fact, well established, and followed closely on the medieval tradition of optics that Alberti had likely studied in Padua and Bologna. It is only when he goes on to expound his theory of perspective that Alberti moves into truly uncharted territory: no medieval text had ever formulated the principles of perspectival drawing, or provided a practical manual for their use.

"Let me tell you what I do when I am painting," he begins conversationally. "First of all, on the surface on which I am going to paint I draw a rectangle . . . which I regard as an open window through which the object to be painted is seen." He then divides the bottom of the painting into equal segments, which he recommends should be one-third of the height of a human figure standing on the "baseline." The next step, he continues, is to pick a "centric point" (which we would call the vanishing point), which can be located anywhere along a line parallel to the baseline at the height of a human figure, known today as the horizon line. "[I]n this way," he explains, "both the viewers and the objects in the painting will seem to be on the same plane."[10] Finally, to complete the construction, he draws straight lines connecting the divisions of the baseline and the centric point.

The resulting pyramid, with its base at the bottom of the painting and its apex at the vanishing point, is, of course, precisely the construction that Brunelleschi had executed, most likely with the aid of a mirror. In the painting's inner space the converging lines appear as parallels that are perpendicular to the painting's surface, point straight toward

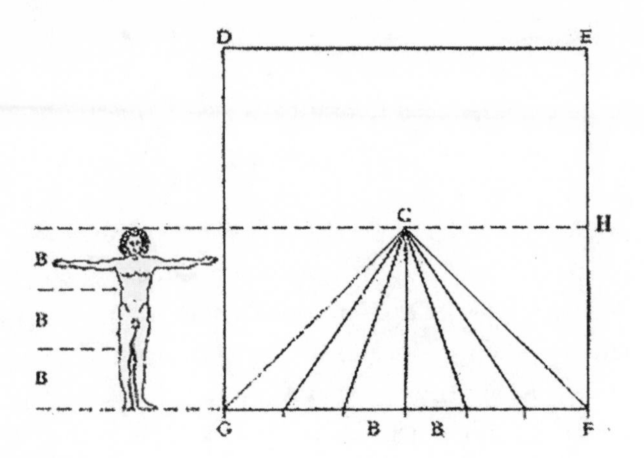

FIGURE 8:
Alberti's depiction of the centric point

the painting's interior, and meet at the vanishing point an infinite distance away. And just as Brunelleschi placed the centric point in his baptistery experiment at the height of his own eye, so does Alberti place the same point at the height of a man in the painting and the height of the observer outside it. In this way the space of the painting appears as a direct extension of the viewer's own space, and the figures in the painting appear about as tall as the viewer.

But Alberti is not done yet. The converging lines show how the distance between objects narrows as they move closer to the vanishing point. But the lines do not show how deep any given point is within the painting's space, and how this depth affects its appearance to a viewer. For that Alberti devised another ingenious method: outside the painting, on its side, he extended the baseline with the same divisions as before. He then marked another point, known today as the distance point, directly above the end of the baseline farthest from the painting, and at the same height as the vanishing point. Finally, he connected the point with the divisions on the original baseline, forming not the symmetric isosceles triangle of the vanishing point but an oblique triangle with its apex at the distance point.

Now comes the tricky part: Alberti draws a line, perpendicular to the baseline, which intersects with the rays connecting the distance point with the divisions of the baseline.[11] He marks the intersection points on the line: the higher up they are (and closer to the horizon line)

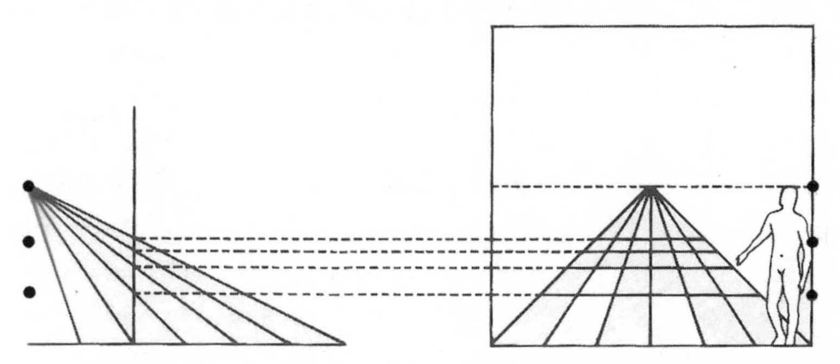

FIGURE 9: **Alberti's construction of the distance point**

the closer the intersections are to each other; the lower down they are (and closer to the baseline) the farther apart they are from each other. Alberti then draws lines through the intersection points and parallel to the baseline, extending them into the framework of the painting itself. There they intersect with the rays connecting the vanishing point and the divisions of the baseline at the bottom of the painting. The result is a grid pattern, broad at the bottom and narrowing as it approaches the vanishing point, like regularly laid tiles extending into the distance. The gap between any two adjacent parallels (those fixed by the distance point) now represents the same magnitude: the seemingly thick "tiles" near the base, which appear almost square, are the same size as the much thinner "tiles" higher up, which appear thin and elongated. In many paintings of the era the squares were, in fact, depicted as actual tiles, and they served as yardsticks for the proper size and shape of objects in different parts of the painting.

This still leaves open an important issue: the rate at which the squares shrink as they recede into the distance is not the same in every case—it depends on the angle of observation. For example, if we are observing a tiled courtyard from the top of a building, then tiles 20 yards away will appear "shallower" than those 10 yards away, but only moderately so. If, however, we observe the same courtyard from normal eye level, then the tiles 20 yards away will seem much shallower than those nearby, and the tiles 30 yards away will hardly seem to have any depth at all. How does one represent different points of view in a painting?

Alberti has an answer: the position of the perpendicular line in figure 2 is not fixed. The line can slide closer to the distance point or farther away from it, thereby changing the points at which it intersects with the rays connecting the distance point to the divisions of the baseline. When the line is close to the distance point, the intersection points toward the top are much closer to one another than the points toward the bottom, which means that the faraway tiles will be much shallower than the ones close by. Conversely, when the perpendicular line is farther away from the distance points, the intersection points are more evenly spread and the tiles will become shallower at a slower rate as they recede into the distance. The first case corresponds to viewing a tiled court from near ground level; the second case corresponds to viewing it from high up.

For the most part Alberti does not explain the geometrical principles behind his procedures, or why they reproduce the magical effect of linear perspective. "I used to demonstrate these things at greater length to my friends with some geometrical explanation," he writes, but here, "as a painter speaking to painters," he decided to omit most of it "for reasons of brevity."[12] Yet the basic principles he is employing are not difficult to grasp: the lines converging on the vanishing point indicate the fixed rate at which horizontal distances between objects shrink as they move into the distance. But it is the gaps between the parallels fixed by the distance point that determine the depth of the picture at every given point. As one draws closer to the vanishing point the depth increases, but not at a fixed rate—at an ever-increasing one. The first construction re-creates the convergence of objects as they move into the distance, like railway tracks receding to the horizon. The second construction decides the depth of the picture at every point along these converging tracks. Taken together they produce a precise geometrical imitation of three-dimensional space as perceived by the human eye.

In *Della Pittura* the patrician Alberti takes Brunelleschi's practical method and transforms it into a systematic theory of perspective. In the baptistery experiment, as we have seen, Brunelleschi had painted an existing structure by faithfully reproducing the perspectival effects he detected in a mirror. The result was a powerful three-dimensional illusion, but it was not altogether clear whether and how it could be applied to other cases. Alberti, in contrast, begins not with a preexisting structure, or even any structure at all, real or imagined. For him it is space itself that comes first: the horizon line and vanishing point, the baseline divisions and the distance point, are all determined before any humans and objects are introduced into the painting.[13] Furthermore, this space, in which the figures will reside, act out their scene, and deliver their moral message, is, for Alberti, entirely geometrical. The space is produced by careful geometrical operations, and its precise shape and depth are determined by geometrical principles. Every point in the painting is defined by geometrical relations, which determine its precise location within the space of the picture and the relative size of an object at that point. The geometry of space will even affect the shape of an object, as we have seen, for example, in the case of a square tile, which will appear nearly square near the base but elongated when near the vanishing point.

The locations of objects in the painting, their relationship and distance from one another, their relative size, and even their shape are no longer determined by the artist's general impressions and sensibility, or by his desire to privilege certain elements over others. The traditional medieval practice of presenting the more important figures, and the most holy ones, in any scene as larger than those less central to the story is now at an end. The space of the painting and everything within it is now fully determined by geometry.

Crucially, geometry is not contained within the artificial space of the painting but extends beyond it, to the natural world. For as Alberti explains in his dedication to Brunelleschi, *Della Pittura* "shows how this noble and beautiful art arises from roots within Nature herself."[14] A painting, if correctly constructed in accordance with the principles of perspective, reflects this all-encompassing geometrical order and replicates it on the canvas. It follows that *Della Pittura* is not a guide for the creation of a closed artificial space that conforms to the principles of geometry. It is, rather, a manual for taking the natural geometric space in which we live and re-creating it on the artificial surface of the painting. For Alberti it is not the painting that is geometrical, it is the world itself.

Nowhere is Alberti's geometric conception of space more in evidence than in his advocacy of the "veil," which, he declares proudly, "I was the first to discover." The purpose of the veil, he explains, is to assist the painter in the process of circumscription—the creation of the geometrical space of the painting:

> A veil loosely woven of fine thread, dyed whatever color you please, divided up by thicker threads into as many parallel square sections as you like, and stretched on a frame. I set this up between the eye and the object to be represented, so that the visual pyramid passes through the loose weave of the veil.[15]

Once this is accomplished the painter reproduces an identical or scaled-down grid on an empty canvas and marks it with the outlines of the objects as they appear through the veil. It is, in effect, a flat surface that shows the outlines of the objects beyond it, which is, according to Alberti, precisely what a painting should be.

The veil is in actual fact what a painting is to be in theory: an intersection of the visual pyramid and a two-dimensional window into a three-dimensional world. Placed between the painter and his object, it immediately reveals the geometrical structure of our lived space. There is the pyramid of visual rays connecting the eye and the observed object, there is the slice of the pyramid—the veil—showing the objects on a flat surface, and there are the regular squares, dividing space itself into regular geometric components that are subsequently filled with objects. Without the veil, the geometrical structure of space is an abstraction, something that we can imagine, and perhaps understand. With the veil geometrical space becomes physical, concrete, and undeniable: it is right there before us. All that remains for the faithful painter is to transfer the geometry of natural space into the imagined space of the painting.

We do not know how many painters actually used the veil in producing their paintings. What we do know, however, is that *Della Pittura* was quickly recognized as the most authoritative treatise on painting of the age, as well as the most useful. Fellow humanists showed an immediate interest in Alberti's theory, so much so that there was soon a movement under way to introduce perspective as a field of study at the universities. Even more significantly, and just as Alberti had intended, painters used it as a technical guide for correct representation. The paintings they produced in the decades after the publication of *Della Pittura* are notable for their rigorous, sometimes even pedantic, adherence to Alberti's methods.[16]

When in 1413 Brunelleschi was conducting his experiments at the Baptistery of St. John he was venturing into new territory. With little if any theoretical grounding, through careful observation and a process of trial and error, he was seeking out the principles of three-dimensional representation. When in the 1420s his protégé Masaccio painted *The Holy Trinity*, *The Tribute Money*, and similar paintings he was applying these practical recipes to the production of remarkable works of art. Masaccio had learned the techniques directly from the secretive Filippo, who undoubtedly made sure to keep them from potential rivals. But with Alberti's *Della Pittura* the principles of linear perspective became more than a series of tricks of the trade, passed on from master to student. They became a well-ordered theory of painting, backed by the authority of a leading scholar and available to any artist.

It was therefore Alberti, rather than Brunelleschi or Masaccio, who was truly responsible for transforming Italian painting. For the generations of artists that followed the appearance of *Della Pittura* the mastery of geometrically correct linear perspective became an absolute necessity, a core competence of any aspiring painter and quite possibly the most important one. Following Alberti's directives, the greatest painters of the age, from Filippo Lippi to Botticelli to Raphael, structured their paintings according to the requirements of geometrical space. Defining and structuring this space was the first step in the execution of any painting, before the figures, structures, and scene were added in.

In previous centuries the painter's primary charge was to present a religious or historical scene in an attractive manner that also made the narrative and moral clear to the viewer. Space entered into the painting only implicitly, as the relative position between the figures and objects; geometry did not enter it at all. But with Alberti the painter's priorities were reversed: universal geometry was ever-present, space structured the painting, and the figures, the scene, the narrative, and the moral lesson would all be subject to its geometrical rule.

The codification of linear perspective in *Della Pittura* was not just a technical innovation: it stood for a profound shift in the understanding of the world humans inhabit. Up to this time the world was characterized by the things within it—houses, people, trees, etc.—and the way they are perceived. Different places seemed and felt qualitatively different and were indeed different from one another, each place with its own unique colors, temperature, texture, and smell. This was very much in line with the dominant Aristotelian philosophy, which taught that things in the world were defined by their perceived qualities: color, texture, and so on. It is also in line with our common day-to-day experience as we interact with our surroundings. Chairs and tables, people and animals, rocks and buildings all exist in this qualitative world. But space itself, that geometrical three-dimensional abstraction? We hardly, if ever, give it a thought.

But beginning in the middle of the fifteenth century, first in Florence and then elsewhere, things change: space is no longer defined by physical objects but becomes an entity in itself—a geometrical construct that we and everything and everyone around us occupy. Now the world is not simply the sum total of what our senses perceive it to be, composed of innumerable different objects with a seemingly unlimited range of color and shape. It is, rather, structured by a deep order, universal and rational, that arranges and pervades all things, regardless of their unique characteristics. The order that structures all is everywhere and always, completely inescapable, universal, timeless, and ubiquitous. It is all around us and yet invisible—until we learn to see it. It is, of course, geometry.

The Secret Code

Despite his emotional return to Florence in 1434, and his declaration that the city was the Albertis' true home, Leon Battista did not settle down in the city on the Arno. As a clerk and secretary in the papal curia he spent the majority of his time in Rome, though this did not prevent him from traveling extensively and visiting the courts of the most glamorous Italian princes. Through it all he never stopped writing, publishing works on the family, on the law, on the soul, on the city of Rome, on ciphers, on statues, and many more; and he never stopped offering his services as an architect and engineer, taking on commissions for the construction and renovation of churches and palaces throughout Italy.

As a result, to Florentines in the later decades of the Quattrocento Alberti was no longer the young man in a hurry he had been in the 1430s but a cultural icon and the pride of the city. A regular visitor to Florence, he took an active part in the lively intellectual life around Lorenzo de Medici (1449–1492), the city's effective ruler from 1469 to his death. Il Magnifico, or "the Magnificent," as Lorenzo was known in even his day, was a patron of the arts on the grandest scale, his court including such luminaries as Leonardo da Vinci, Sandro Botticelli, and Michelangelo. A dedicated humanist devoted to the recovery of ancient philosophy, he sponsored the Platonic Academy and supported its illustrious members, including Marsilio Ficino and Pico della Mirandola. Under his guidance Florence was, without question, the artistic and intellectual leader not only of Italy, but of all of Europe.

To Lorenzo and his circle Alberti was a heroic figure, a member of an older generation who seemed to them a living embodiment of the humanist ideal. A scholar of the highest accomplishments, Alberti was also a practical man of the world, a great admirer of the ancients who did not hesitate to put their teachings to the test of the here and now. Classicist, philosopher, moralist, author, and architect, he was all that Il Magnifico and his followers aspired to be, and when they visited Rome in 1471 to pay their respects to Pope Sixtus IV they would have no other guide to the mysteries of the eternal city than their admired and beloved compatriot.[17] His prestige and authority among them was boundless: as Lorenzo's friend the philosopher Angelo Poliziano put it, Alberti "was a man of rare brilliance, acute judgement, and extensive learning . . . Surely there was no field of knowledge however remote, no discipline however arcane, that escaped his attention."[18]

. . .

THERE WAS, HOWEVER, NO FIELD of knowledge to which Alberti was more dedicated, or in which his authority was more absolute, than architecture. In the early 1440s, following in the footsteps of Brunelleschi, he conducted his own survey of the ancient ruins of Rome, searching for the hidden mathematical harmonies that guided the ancient Romans. The secretive Brunelleschi had kept his discoveries to himself, but the sociable and convivial Alberti recorded all his findings in great detail and cited them extensively in his architectural writings. True to his geometricizing spirit, he placed the contours of the walls, the gates, and the great buildings and monuments of the city, both ancient and modern, on a precisely measured grid in the shape of an astrolabe and published the coordinates in a short treatise entitled *Descriptio Urbis Romae* (*Description of the City of Rome*).[19] And so, when the forward-looking pontiff Nicholas V (1447–1455) launched a building program aimed at restoring Rome's greatness, it was only natural that he turn to Alberti, his greatest architect, to serve as his chief consultant.[20]

In 1452 Alberti presented Nicholas V with a book that he considered his greatest work, and it was, appropriately, a treatise on architecture. Entitled *De Re Aedificatoria* (*On the Art of Building*), the book was written in the classical Latin favored by humanists, and its ten books were ostensibly a commentary on an ancient work: the ten books of Marcus Vitruvius Pollio's *De Architectura*, written in the first century B.C.E. and recovered by Poggio Bracciolini in 1414. This was to be expected, since Vitruvius's text was by far the most comprehensive discussion of the art of building known to Alberti's generation. Nevertheless, as Alberti and his contemporaries knew well, *De Re Aedificatoria* was much more than a commentary on an ancient text: it was a bold statement of a novel theory of building unlike any that came before it. It may be less known today than the far more readable and charming *Della Pittura*. Yet no text did more to shape the urban landscape of the Renaissance—and through it, of modern cities.

To Il Magnifico and his circle the ten books of *De Re Aedificatoria* were something close to architectural scripture. When in 1485 it became the first treatise on architecture ever to be printed, Lorenzo ordered a copy directly from the printer and waited impatiently for the book to arrive. Throughout 1486, as the folios of *De Re Aedificatoria* trickled in (printers at the time were not responsible for binding), he had them read before his friends and companions, page by page and chapter by chapter.

The contrast between *De Re Aedificatoria*, Alberti's mature masterpiece, and *Della Pittura*, written seventeen years prior, is striking. *Della Pittura*, as we have seen, was a short, accessible, and practical manual, written in Italian and meant for the use of working artists. It was also one of Alberti's early works, and he no doubt would have been surprised, perhaps even shocked, to learn that it was this unpretentious little booklet that secured his fame for centuries to come. *De Re Aedificatoria*, in contrast, was the ambitious work of a famous scholar, seeking to lay claim to the intellectual leadership of his generation. Far longer and written in a florid and difficult Latin, it was aimed at fellow scholars, the kind of people who congregated at Lorenzo de Medici's table.

Yet despite their differences, a single vision ran through both works: that the world was fundamentally geometrical and could therefore be described in accordance with geometrical principles. In *Della Pittura*, the theory of perspective was the reproduction of this inherent geometry on a painter's canvas. Similarly, in *De Re Aedificatoria*, the well-designed mansion and garden followed this deep geometrical order. A painting, structure, or landscape designed in accordance with the true principles of geometry was therefore beautiful. One that was not so designed was deformed, or simply ugly.

It followed that for Alberti beauty could not be simply in the eye of the beholder: for if one person saw beauty in what another considered ugly, there would be nothing left of the universal standard of beauty he was seeking. And so, while he allowed for a small degree of variation in opinion, Alberti insisted that we all share the same core standards of beauty. "That a thing is beautiful," he wrote, "does not proceed from mere Opinion, but from a secret argument and Discourse implanted in the Mind itself." To discover the rules of architecture, in other words, one must unveil the natural geometrical order that makes things beautiful.[21]

Setting out to unveil the hidden geometrical code of nature, Alberti began *De Re Aedificatoria* with the simplest and most perfect geometrical figure of all—the circle. "It is manifest," he wrote, "that Nature delights principally in round figures," citing "stars, trees, animals, the Nests of Birds." It follows, he continues, that the area for a round temple should be "exactly circular." Hexagons are also favored by nature, he notes, as is the case with the cells in the hives of bees, hornets, and wasps, but this does not mean that all properly designed structures should be hexagons. It is enough that they be symmetrical and precisely delineated polygons of "six, eight, or sometimes ten sides," and that they adhere to certain measurements.[22]

Alberti's directions left nothing to chance. In a quadrangular design, he decreed, the length should exceed the breadth by one half or one third. If these proportions are applied more than once, one might arrive at additional possibilities: for example, the length of a platform with a breadth of 4 might exceed it by one-half, making 6. But if this same proportion is applied again the length becomes 9, and the proportions of the platform would be 4 to 9. Similarly, if the length exceeds the breadth by one-third, applying the proportion twice over to a 9-foot-wide rectangle will result in a hall with a breadth-to-length proportion of 9 to 16. Alternately, applying the two proportions twice or three times can also produce surfaces of which the length is double, triple, and quadruple the breadth. All of these proportions are, to Alberti, "natural," and therefore beautiful.

If one prefers to build a structure with more than four sides, then it should always be shaped as a perfect regular polygon, with equal sides and angles. Such polygons are, in effect, approximations of the most "natural" figure, the circle, and can be derived from it through proper geometrical construction. To create such ground plans, Alberti explains, one must start with a circle and follow some strict rules of geometry. To create a hexagon, for example, one should take the radius of a circle and place it so that it touches the circumference at each end. Repeating the process, with the radii placed around the circle end to end, will produce a perfect hexagon enclosed within the circle. Producing regular polygons of eight, ten, and twelve sides, Alberti shows, is only slightly more difficult.[23]

Everything in an architectural design, according to Alberti, must conform to proper geometrical proportions. For example, "if the Length of the Platform be twice its Breadth; then, where the roof is to be flat, the Height must be equal to the Breadth; where the Roof is to be vaulted, a third part of that breadth more must be added."[24] Things got only more complicated from there: if the length of a structure is four times the breadth, the height should be half the length for a vaulted roof, three-quarters the breadth for a flat one; but if the ratio of length to breadth is five to one, then one should increase those figures by one-sixth of the height. And so it goes. The width of the walls too should conform to strict proportions: those separating one chapel from another "should never have less Thickness than the fifth part of the Break that is left between them, nor more than the third."[25] But if the central design is round, surrounded by six chapels, then the thickness of their walls should be one-half the distance between the chapels. Even the three ancient architectural orders, the Doric, Ionic, and Corinthian, he argues, are not arbitrary: each is defined by precise geometrical proportions, derived directly from the natural proportions of the human body.[26]

In page after page of *De Re Aedificatoria* Alberti lists the exact geometrical proportions to which a beautiful design should adhere. To us, looking back across the centuries, such prescriptions seem entirely arbitrary. Why these particular proportions and not others? In fact, why have such rigid proportions at all? Isn't something beautiful simply because we consider it beautiful, regardless of whether it is regular, symmetrical, and geometrical? And does it really matter if the height of a rectangular hall whose length is twice its breadth is exactly the same as its breadth? Alberti thought that it did. These geometrical proportions, he had no doubt, were not arbitrary or whimsical, but expressions of the deepest harmonies that govern the world around us. Many of them, he believed, were known to the ancient Greeks and Romans and can be discovered by measuring and surveying the structures they left behind, as he and Brunelleschi had both done. But at their core, they are the measures not of antiquity but of nature herself.

For Alberti, in other words, the role of architecture was to bring out the deep geometrical order that governs everything in the universe. That order is, to be sure, all around us, in everything from the shape of a bird's nest to the dimensions of the human body, structuring and ordering the world. But it is hard to detect in the overwhelming richness and seemingly infinite variety of the world we sense and see. It is therefore left to the architect to make this hidden geometry visible in his creations. The circles, rectangles, and regular polygons that are, according to Alberti, the building blocks of any architectural design are not human impositions on the natural landscape. They are, rather, expressions of a hidden but nevertheless real and all-pervading natural order. The truer an edifice is to this universal geometry, the truer it is to the science of architecture and the more beautiful it is in the eyes of men.

In *Della Pittura* Alberti taught practicing and aspiring artists how to encode the geometry of the world in their paintings, so as to produce beautiful works of art that were also true to nature. Two decades later, in *De Re Aedificatoria*, he taught architects how to reveal the same hidden geometry by constructing buildings within the natural landscape itself. And so, despite their differences and the fact they were written at opposite ends of a long career, the two works are, in a way, twins. Both are inspired by Alberti's thrilling discovery that the world is geometrical; and both are aimed at spreading the word and adapting human creations to this newfound reality. Alberti seemed to acknowledge this connection when he wrote in *De Re Aedificatoria* that "[t]he Arts which are useful, and indeed absolutely necessary to the Architect, are Painting and Mathematicks." The law, astronomy, rhetoric, or any other discipline might prove useful to an architect in some circumstances, but are hardly essential. "[B]ut Painting and Mathematicks are what he can no more be without than a Poet can be without knowledge of Feet or Syllables."[27] For painting, just like architecture, is founded on the mathematical laws of geometry.

The rules of perspective, which Alberti set out so clearly in *Della Pittura*, have been enshrined as a great discovery that changed the course not only of art but also science. Quite apart from revolutionizing painting from that day to our own, linear perspective also brought about a deeper understanding of optics, leading to the inventions of the telescope and the microscope more than a century later. The same cannot be said of the principles of architecture laid out in *De Re Aedificatoria*. Unlike the rules of perspective, these strike us not as discoveries about nature, but as arbitrary impositions on free creativity. No modern architects would consider themselves bound by Alberti's rigid decrees.

Yet in Alberti's own lifetime, and for many generations thereafter, judgments were different. To say that *De Re Aedificatoria* was an influential book in its time would be a huge understatement. It is more accurate to say that it was the Bible of architectural teachings, and that it defined the standards for classical architecture up to the dawn of the nineteenth century. For the fact is that the skylines of Europe and North America would have been very different if not for the Florentine polymath's powerful intervention. From St. Peter's Basilica in Rome to St. Paul's Cathedral in London and on to the Panthéon in Paris and the Capitol Building in Washington, the teachings of *De Re Aedificatoria* have fashioned the most monumental structures of Western capitals. For all of them are guided by what Alberti considered his greatest discovery: that nature is geometrical through and through, and human art must become so as well.

The Geometrical World

On the south side of the Piazza del Duomo in Florence, about halfway between the cathedral and the Baptistery of St. John, is an ornate two-story columned building known as the Loggia del Bigallo. Built as a monastery in the middle of the fourteenth century, the Loggia is home to a fresco from 1352 depicting the Madonna della Misericordia (the Virgin of Mercy) towering over a crowd of the adoring faithful. The painting is not unusual for its time, except for one feature: at the bottom of the fresco, just below the Madonna's feet, is the oldest known map of the city of Florence. There at the center is the Baptistery of St. John, the pride of the city before Brunelleschi completed the dome of the cathedral nearly a century later. Also recognizable is the famed square tower of the Palazzo della Signoria, hemmed in from all sides by roofs and buildings. The rest of the city is a jumble of towers, churches, and palazzos,

FIGURE 10: **Map of Florence at the Loggia del Bigallo (1352)**

all packed within the city walls and crowding in on one another in a tight, bewildering mass.

The depiction of Florence is not unrealistic: in fact, the overall effect is very much what one experienced in the narrow streets and crowded alleys of a medieval city, with a church around every corner and massive stone edifices looming on every side. It has an immediacy and a feel for the city that is entirely absent from modern-day maps. What the Bigallo map lacks, however, is any sense of a preexisting orderly space in which the structures are located. The buildings that make up the city are simply there, piled up on top of one another without any suggestion of depth, distance, or location. Nearly a century before Alberti published *Della Pittura*, the geometrical space that we take for granted simply wasn't there.

Now compare this fresco to another map of Florence. Created around 1485 by the painter and mapmaker Francesco Rosselli, it is known to us through a later copy as the Map with Chain. Florence had not

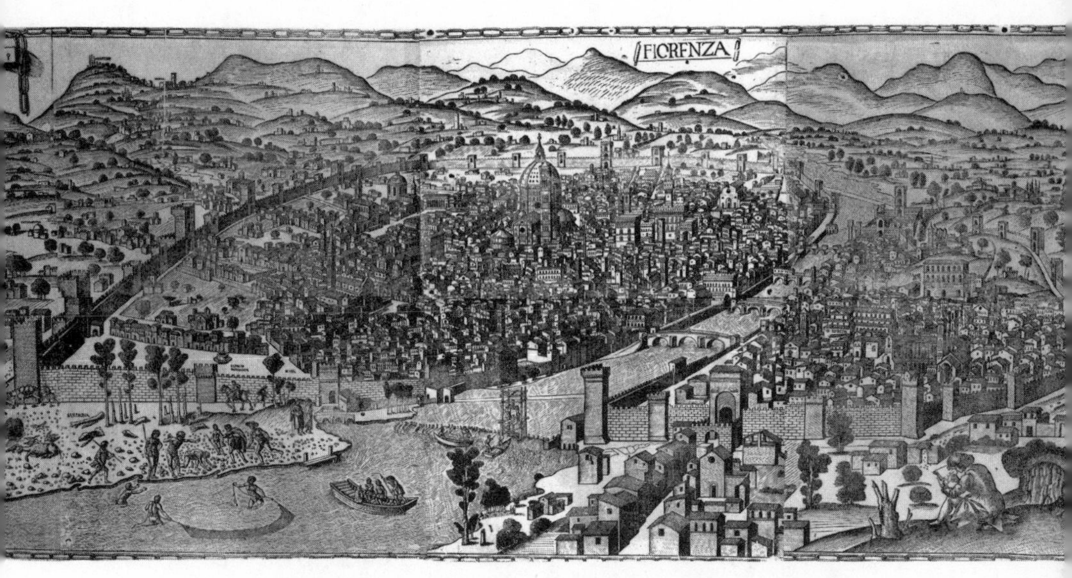

FIGURE 11: Map of Florence known as the Map with Chain. Woodcut by Lucantonio Degli Uberti, after Francesco Rosselli, ca. 1485

changed much since the artist of the Loggia del Bigallo completed his work, with the exception that the grandest structure in the city now was no longer the baptistery, but the cathedral with its great dome. Yet if the city was much the same, its depiction could hardly be more different. For one thing, whereas the relative size of the churches and palazzos in the older map was entirely impressionistic, reflecting their significance in the eyes of the artist rather than physical size, Rosselli is careful to represent the correct sizes of the main edifices, showing them towering over the neighboring buildings. But more critically, the Map with Chain, unlike the map in the Loggia, is executed strictly in accordance with Alberti's rules of perspective.

Whereas the Loggia fresco depicted each building as seen close-up, Rosselli's map presents us with a clear view of the city from a particular vantage point, a bird's-eye view from the southwest. Following Alberti's guidelines, the walls on the right and left converge to a vanishing point beyond the city, and the broad river in the foreground, as well as the rapidly shortening vertical distances as the map stretches through the city toward the hills, point to the systematic application of a distance point. As a result, every church, tower, palazzo, and dwelling is placed in its own precise location within the overall space of the map. In the Loggia map, space is an afterthought, a mere byproduct of the jumble of structures that make up the city. In the Map with Chain, space comes first, rigorously defined by the laws of geometry, and the city and all its structures live within it. The Loggia map from the 1350s and the Map with Chain from the 1480s depict the same city. But they also depict different worlds.[28]

SOMETHING DEEP AND IRREVOCABLE HAD happened during the century that separates the two maps. The map at the Loggia exists in the qualitative, impressionistic world of the Middle Ages, in which things were experienced and depicted in accordance with their size, texture, and spiritual or political significance. It is a world given philosophical expression in Aristotelian physics, in which objects were defined by qualities such as color, texture, and temperature. The Map with Chain occupies an entirely different world, defined and governed by the strict rational laws of Euclidean geometry. It points the way to modern science, which seeks out the hidden mathematical order deep within all things. The Loggia map is a mature expression of the great medieval civilization, whose highest artistic, architectural, and intellectual achievements date from those very years. The Map with Chain comes at the earliest stirrings of modernity. It is a glimpse of things to come.

The chasm that separates those two worlds can be traced with stunning precision. Brunelleschi, Masaccio, and Alberti lived in the same place at the same time and knew one another. They were members of a single tight-knit community of artists, craftsmen, and architects who walked the same streets of a single Italian city—tiny by modern standards—in the early decades of the fifteenth century. Good humanists that they were, they thought they were reviving ancient glories. In fact, they were ushering in an unprecedented new world, one we call modern. For in Florence in the 1420s and '30s Brunelleschi, Masaccio, and Alberti made the world geometrical.

PART II

Euclid's Kingdom

3.

ROYAL GEOMETRIES

The Italian Obsession

King Charles VIII of France (1470–1498) stood on the ground floor of the royal villa of Poggio Reale, a few miles outside Naples, looking down at the garden below. It was March of 1495, and the young king was not well. As he explained in a letter to his relative, the cardinal of Bourbon, he had contracted the measles, and had retired to the country villa to recover his strength.[1] The treatment seemed to be working, as the monarch slowly recuperated from what at the time was a life-threatening illness. Charmed by the regular geometrical designs of the garden's parterres, a determination formed in the king's mind: he would re-create the patterns laid out before him at his own royal castles in France.

Charles's decision, though rarely mentioned at the time, may well have been the most consequential of his fifteen-year reign. For there, in the hinterland of southern Italy, a link that would only grow tighter with the passage of time was forged between the kings of France and their geometrical gardens. Over the next three centuries gardens would become the regime's most brilliant emblems, unmatched founts of prestige and authority. From Charles VIII to Louis XVI, and from the Tuileries to Versailles, nothing bespoke royal power in France like geometrical gardens. In the minds of both kings and their subjects, the two became well-nigh inseparable.

Back in 1495 Charles could not envision the brilliant geometrical gardens his successors on the throne of France would create and cherish. Wherever he turned his gaze, all he could see were the slopes of Mount Vesuvius and the gently rolling hills of the Neapolitan countryside, prizes of his brilliant military success. It had been only seven months since the king had led his armies across the Alps and onto the plains of northern Italy. Charles was out for revenge: fifty-two years had passed since René of Anjou, the last French king of Naples, lost his kingdom to Alfonso V of Aragon (thenceforth also Alfonso I of Naples), fifty-two years in which the kings of France, rightful heirs to the Angevin (i.e., Anjou) line, had threatened repeatedly to invade the Italian peninsula and restore French honor by force of arms. Fortunately for the Aragonese, not to mention the Italians, the French monarchs in those days had their hands full and never made good on their threats. Charles VII (1403–1461), who was crowned in Reims in 1429 under the beatific gaze of Joan of Arc, spent most of his reign in a desperate and ultimately victorious fight against the English and the Burgundians, who controlled much of France. His son and successor, Louis XI (1423–1483), took decades to dismantle the mighty Duchy of Burgundy and set the French crown on the long road to recovery from the disasters of the Hundred Years' War.

And so it was left to Louis's son, Charles VIII, to uphold the French claim to the kingdom of Naples. By 1494 Louis was no longer the thirteen-year-old child who had ascended the throne more than a decade before, but, in his own eyes at least, a shining emblem of chivalry and leader of Christendom. Raised to view himself as a worthy heir to his namesake, Charlemagne, he considered the Italian expedition the first step toward reclaiming his forebear's great empire and leading a crusade to liberate Constantinople from the Ottoman Turks.

Others, however, were more skeptical of the king's ambitions: "The King of France," wrote Contarini, the Venetian envoy to the French court, is "small and ill shaped, facially ugly, his eyes are large and pale and inclined to see badly rather than well . . . His hand is convulsed by spasms which are very ugly to see." Charles's intellectual gifts, in the Venetian's opinion, did nothing to compensate for these deficiencies: "He speaks slowly," the ambassador noted, and "I believe that in body and mind he is of little worth." Feeling perhaps obliged to soften this harsh assessment, Contarini noted the king's skill in tennis, hunting, and jousting, "activities to which," he added dubiously, "for good or ill, he devotes much time."[2]

But if the king himself was less than prepossessing, his army was a different matter. For centuries French monarchs had been served by a feudal army, mustered by the great nobles of the land from among their dependents and placed under the king's command in times of war. Large, imposing, magnificently attired, and seemingly invincible, the royal army nevertheless proved disastrously ineffectual when faced with the much smaller but highly disciplined armies fielded by the kings of England. In battle after battle during the Hundred Years' War, the proud charge of the French knights withered when faced with a hail of arrows launched from thousands of English longbows. The humiliations of Crécy (1346) and Poitiers (1356) in the early years of the war culminated in 1415 with the wholesale slaughter of the flower of French nobility in the battle of Agincourt. At long last learning from this bitter experience, the French kings finally gave up on their feudal military system and looked for alternatives. Hesitantly at first, but then with increasing confidence and success, they established a standing army that was paid from the royal coffers and answered directly to the king, rather than to his independent-minded nobles. With this new force at their disposal the French kings turned the tide on the English invaders and their Burgundian allies. They defeated them in countless skirmishes and several sizable battles, confined them to castles and fortified towns, and then captured these systematically and efficiently with the aid of their advanced artillery. By 1483, when Louis XI breathed his last, the English had been expelled, the Burgundians and other great nobles brought to heel, and royal rule extended over nearly all the territory of modern France.

Battle-hardened, experienced, and efficient, this was the army that Charles VIII led over the Alps in the spring of 1494. Close to half of the 27,000-strong force was made up of traditional knights, cavalrymen bearing lances and heavy suits of armor who were little different from their forebears who had charged the English defenses at Agincourt eighty years before.[3] This was to be expected from a monarch who was himself a knight and still relied on the provincial nobles to extend and legitimize his rule. The strength of the army, however, lay elsewhere: the core of the infantry was composed of Swiss pikemen, the most feared mercenaries in Europe. Fighting on foot in tight formations reminiscent of the Macedonian phalanx, the Swiss had repeatedly bested the armored knights sent against them by their far larger and more powerful neighbors. Alongside them were large numbers of French crossbowmen and archers, who provided both cavalry and infantry with the support and cover so sorely lacking in the early decades of the Hundred Years' War.

Yet it was not the knights in their shining armor or the disciplined infantry units with their pikes and bows that most impressed contemporaries. It was the artillery train that followed them on their march. Cannons, to be sure, were not new: they had been known in Europe for over a century and had been used regularly in reducing fortifications. These "bombards," as they were known, were cumbersome instruments, assembled specifically for a siege, practically immovable, and capable of firing off only a few shots a day. The artillery of Charles VIII's army was something else altogether. Thanks to improved methods of bronze casting, and a more powerful propellant known as "black powder," French cannons became far more powerful, and capable of firing solid shot every few minutes instead of every several hours. Just as important, the cannons were lighter and could be mounted on wheels and pulled by horses. In a battle or siege this meant that their aim could be adjusted and they could be pointed in different directions. And when the army marched, unlike the bombards of old, they followed right along with it.[4]

The arrival of the French royal troops on the plains of Lombardy shook the Italian city-states to the core. Regardless of whether they supported Charles's claim to the throne of Naples, one thing was clear: no Italian prince or city possessed the population, resources, or technology to compete with the might of Charles's army. The old rivalries between Florence and Milan, Venice and Naples, which had defined the Italian political landscape for centuries, seemed suddenly insignificant in the face of this new challenge; as long as the French king was loose on the peninsula, no one was safe.

The city-states of Italy were no strangers to armed conflict: rival city-states had fought one another for centuries, and their citizens considered war to be the normal state of affairs between neighboring cities. Yet what the Italians considered war and what the French brought along with them were entirely different things. By the fifteenth century the princes and republics of the peninsula, wary of arming their restive populace, had disbanded their citizens' militias and contracted all fighting duties to bands of mercenaries known as condottieri. Professionals out to make a living, they were only loyal as long as they were paid, and were ever ready to abandon their employers, or even switch sides, if a better deal was in the offing. They had little interest in throwing away their lives for a cause they cared nothing about and never lost sight of the fact that today's enemy could well be tomorrow's employer.

As a result, battles between bands of condottieri tended to be relatively mild and predictable affairs, in which victory was granted to the larger and better-paid band. The notorious Florentine statesman Niccolò Machiavelli (1469–1527) was probably exaggerating when he claimed that in the day-long battle of Anghiari (1440), fought between Florence and Milan, "only one man was killed," and only because he "fell from his horse and was trampled to death." But there is no question that compared to the horrors taking place in northern Europe, Italian warfare in the Quattrocento was a relatively bloodless affair.[5]

Charles's army put an end to all that, sweeping through Italy like an irresistible tide of brutality and efficiency. When in October of 1494 the Neapolitans sought to preempt the French advance by occupying the strategic castle of Mordano, they probably expected to hold out for months or years, as would have likely been the case if their enemies had been fellow Italians. But French artillery breached the walls in a mere three hours, and the French troops that stormed the castle massacred all its occupants. When the Neapolitans tried again at the fortress of Monte San Giovanni the results were the same, as was the massacre that ensued. One by one the Italian city-states rushed to spare themselves by making peace with the conqueror. In Florence, Piero de Medici, the city's effective ruler, was chased from the city as the French army stood at the gates, and replaced by a republican government friendly to the invader. In Naples, King Alfonso II abdicated his throne and entered a monastery, and his successor, Ferrante II, fled for his life. On February 22, 1495, Charles entered the city beneath a canopy of gold cloth borne by four Neapolitan noblemen and declared himself the legitimate king of Naples.

In the end, Charles's victory turned out to be short-lived. His triumphant march from Lombardy to Naples rattled not only the Italian city-states but also the great princes of Europe, including Pope Alexander VI, Emperor Maximilian I of the Holy Roman Empire, and Ferdinand II of Aragon, who now banded together with Venice and Milan in the League of Venice. Three months after his sojourn in Poggio Reale, Charles met the League's combined forces in battle near the town of Fornovo in the first serious challenge the French army had faced in the campaign. Despite being greatly outnumbered, Charles's veterans once again proved their worth, holding off their enemies and securing their lines of retreat. Yet the French king's Italian dream was at an end: his army marched back across the Alps, and by 1496 Ferrante II was back in on the throne of Naples. When Charles died only two years later, any plans he might have hatched for a new expedition died with him.

By the end of the century an outside observer might conclude that despite the dramatic events of 1494–1495, the old order had survived and was safely restored to Italy. The coming years would prove otherwise: Charles's swift march through the peninsula had exposed the bankruptcy of the Italian political system in a way that could no longer be covered up. The Italian city-states were populous, cultured, and fabulously wealthy. They were the artistic and intellectual teachers of Europe, and key players in international banking and commerce across the continent and beyond. But they were also, it turned out, defenseless against a determined assault by their larger and more powerful neighbors. Now that their vulnerability had been exposed, they were to prove an irresistible temptation to the great national monarchies then coming into being, especially in France and Spain.

As long as the Italian powers stuck together in a common defense they could usually deter foreigners from interfering in their affairs. But whenever one of the endemic intrapeninsular quarrels broke out—between the Papacy and Florence, or between Milan and Naples—it was almost inevitable that one side or the other, or sometimes both, would call upon foreign princes for help. As a result, for the next hundred years Italy became the favorite battleground for France and Spain, the place where the rival kingdoms tested their strength against each other and vied for supremacy in Europe. Time after time French kings marching down from the north would be met in battle by Spanish monarchs coming in from the south in a war that produced no clear winner. The incontrovertible losers, however, were the people of Italy, who saw their lands ransacked, their property looted, and their families and fellow citizens abused and sometimes massacred by marauding foreign armies. Charles VIII's lightning campaign, it turned out, was only the opening chapter in what would prove to be a century of woe for Italy.

A Villa in Campania

But as the triumphant king stood atop the garden at Poggio Reale in the early spring of 1495, he most likely was thinking not of the ravages of war, but of the blessings of peace. Having led his troops to one decisive victory after another, and suffered bravely through the rigors of the campaign, Charles was now ready to enjoy the fruits of his success by retiring to comforts worthy of his royal station. Poggio Reale, a favored retreat of the Aragonese kings of Naples, seemed like the perfect place to recuperate from his illness and recover his strength. Built by

Alfonso II in the 1480s, the villa was already famous in its day not only for the elegant square house in the classical style at its center, but also for the surrounding gardens. "You would not credit the fine gardens I have in this town," the enchanted king wrote to the cardinal of Bourbon. "For by my faith only Adam and Eve seem lacking to make them an earthly Paradise, so good and so beautiful are they, and so full of remarkable things."[6]

Despite Charles's reputation for frivolity, his fascination with the gardens he encountered in Italy proved serious and enduring. Much to the king's consternation, the royal baggage train, including all the trophies and works of art Charles had collected and looted during his campaign, was captured by the League's forces in the battle of Fornovo. Yet the king's two most important Italian acquisitions remained in his possession and returned to France with his army. These were not the beautiful paintings and sculptures we associate with the Italian Renaissance, nor were they hoards of gold and silver captured from his Neapolitan enemies. They were, rather, two master gardeners who, voluntarily or not, entered into Charles's employ, joined his entourage, and followed him to France. They had only one task: to work the magic the king had witnessed in Poggio Reale and create gardens of Eden for the king's own castles and palaces beyond the Alps. Remarkably, of all the changes—military, political, cultural—brought about by Charles's invasion of Italy, it was this little-noted outcome that was to prove most lasting and most consequential. For it is not a stretch to say that the seeds of Versailles, brilliant emblem of Louis XIV's absolute monarchy, were sown in the garden of Poggio Reale.

What did Charles see in the garden of that royal villa outside Naples that so captivated him? We do not know any specifics, as not a trace is left today of either the house or the gardens of Poggio Reale. A reconstruction by the Dutch engraver Bastiaen Stopendael shows a handsome square structure with towers at the corners, surrounded on three sides by a walled garden that is divided into several rectangular segments and includes a large rectangular pool. But the engraving dates from the late 1600s, a full two centuries after Charles's visit. By that time the villa had been extensively modified, and had in any case long been surpassed by far grander princely estates. There is nothing in Stopendael's engraving that explains the deep fascination the villa and its garden held for the king and his companions.[7] For that we need to turn elsewhere.

Although the architectural details of Poggio Reale are forever lost to us, there is much we nonetheless do know about the kind of villa and the kind of garden that they were. In its time it was one of the first and most ambitious of a new kind of estate that was gaining currency in Renaissance Italy. These new princely enclaves reflected a novel understanding of space, and of the world as a whole: unlike their medieval predecessors, estates such as Poggio Reale were built to show that the entire world, and space itself, were geometrical through and through.

Building the Earthly Paradise

At the time of Charles's visit, princely gardens were a relative novelty in Western Europe. Throughout the Middle Ages, inasmuch as gardens existed at all, they most often were part of monasteries, not noble courts. Far from being sites of enjoyment and relaxation, monastic gardens were entirely utilitarian, commonly including a vegetable patch, medicinal herbs, an area set apart for fruit trees, and a fishpond. Planted and tended to by the monks, the garden would provide the brothers with a reliable supply of fresh food for their table. Surpluses could be traded for other useful goods from the neighboring villages, thereby enhancing the monastic and rural economies.[8]

Secular gardens were much rarer in that age, but with them too, utility was the guiding principle. In a period of endemic warfare, some great lords thought it in their interest to attach gardens to their castles. As depicted in the Duke of Berry's Book of Hours, such gardens were far larger than the monastic ones, sometimes including open meadows, woods, and lakes.[9] For the most part they served as a source of fresh food for the castle's residents, but also, as the Book of Hours illustrates, as a place where great lords and ladies could socialize in their leisure time.[10]

The growing social function of gardens can be glimpsed in the *Roman de la Rose*, dating from the thirteenth century and likely the most popular chivalric romance of the high Middle Ages. The tale of the poet's pursuit of the beautiful rose takes place in a garden filled with dangers as well as delights. The biblical tale of the Garden of Eden, with its fill of wonders and temptations, and the "closed garden" of the *Song of Songs* could not have been far from the medieval poet's mind. Yet the *Roman de la Rose*'s descriptions of the trees, fruit orchards, and flower beds suggest that it is based not only on the poet's religious musings and inspired imagination, but also on actual gardens that he had seen and experienced at first hand.

Rare as they were, such medieval pleasure gardens were small and simple, dominated by open, unstructured "flowery meads" with fruit trees scattered here and there. It was not until the publication of the Italian scholar Pietro de' Crescenzi's *Opus Ruralium Commodorum* in the fourteenth century that the reality of pleasure gardens began to catch up with the imagined ideal in the *Roman de la Rose*. The eighth book in de' Crescenzi's treatise was devoted specifically to pleasure gardens, whose only purpose was the enjoyment of their owners. These, according to de' Crescenzi, should be carefully designed spaces, enclosed by a wall or hedges, and arranged according to a systematic, overarching plan. In addition to the traditional orchard they should also have fountains of running water, a fishpond, an aviary for exotic birds, and—if large enough—a park for wild animals. At the center of it all would be the great house or palace, and everything, from the orchards to the animals in the park, should be viewable from its windows.

"Orchards of Royalty," de' Crescenzi called such gardens, and little wonder, for only kings or the greatest of lords possessed the lands, the resources, and the time to maintain and enjoy them. Conversely, having a pleasure garden on one's estate bestowed immense, even royal prestige on its owner. Charles V of France (1338–1380) certainly took notice, for he not only ordered de' Crescenzi's treatise to be translated into French, but also created what were likely the first royal gardens in France, at the Hôtel Saint-Pol in Paris. As an early effort to associate the power of the French monarchy with gardens, the green pastures of Saint-Pol were a sign of things to come, but they did not outlive their master.[11] It may have been due to the disastrous turn in French fortunes in the Hundred Years' War, or it may have been a matter of personal preference, but the kings who succeeded Charles V showed no interest in cultivating royal gardens. It was not until Charles VIII's fateful stay at Poggio Reale that pleasure gardens once again became a concern of the kings of France.

Although Poggio Reale was situated just a few miles outside of Naples, the villa's true origins were in a different Italian metropolis, hundreds of miles to the north: the Florence of Lorenzo de Medici. Under Il Magnifico's guidance Florence had become the undisputed master of Europe in both the letters and the arts, and when Ferrante I of Naples sought to build a new royal villa outside his city's walls, it seemed only natural that he would turn to Lorenzo, his friend and ally, for advice. In 1487, almost certainly at Lorenzo's instigation, two models of the villa were sent down from Florence to Naples along with Giuliano da Maiano (1432–1490), a Florentine architect and member of Lorenzo's inner circle. Maiano spent the next three years working at Poggio Reale, and it is widely believed that he is chiefly responsible for the design of the villa and gardens that so captivated Charles VIII. When Maiano died in 1490 with the project still ongoing, Lorenzo sent down another of his court architects, Luca Fancelli (1430–1495), to succeed him.[12] The two masters, each in turn, presented the Neapolitan king with a villa and garden that expressed their highest ideals of beauty and functionality.

Maiano and Fancelli were humanists as well as architects, and members of the tight-knit group of scholars gathered around Il Magnifico. They were almost certainly present when the pages of Alberti's *De Re Aedificatoria*, fresh from the printing press, were read aloud at Lorenzo's table in 1486. To them, as for other members of Lorenzo's circle, Alberti's opus was architectural gospel, obsessively discussed and enthusiastically embraced. When in the very next year Maiano was dispatched to Naples to work on Poggio Reale, and when Fancelli followed him three years later, the teachings of their late hero, Leon Battista Alberti, were fresh in their mind.[13] And so, while we do not know the precise outlines or measurements of the famous garden at Poggio Reale, we do know the key principles that guided its design: it was designed to be an Albertian garden, an ideal representation of the principles of *De*

Re Aedificatoria. This means that Poggio Reale was first and foremost a geometrical garden.

We have already seen how Florentine scholars and artists of the Quattrocento made the world geometrical. Long admired for its absolute truth and certainty, the science of geometry was previously thought unsuitable for describing the irregular and transitory physical world. But by revealing the laws of linear perspective, Brunelleschi, Masaccio, and Alberti changed that, demonstrating that the seemingly limitless variety one encounters in nature is in fact governed by the fixed, eternal laws of geometry. In addition to changing Europeans' understanding of the physical world, the discovery also recast the human arts such as painting and architecture. Henceforth their purpose was to reproduce the hidden geometrical order of the world, and thereby bring it to light. As Alberti explained in *De Re Aedificatoria*, the physical world is composed of circles, rectangles, and regular polygons, which structure everything from the human body to the shape of a flower or a beehive. A properly designed building, it followed, will make use of these same geometrical shapes and proportions, thereby re-creating both the functionality and the beauty of the natural world. Consequently *De Re Aedificatoria* is, for the most part, a detailed manual on how to reproduce the geometrical harmonies of nature in the design of churches and palazzos

But what of gardens? Those had been integral to the design of the ancient Roman villas that Alberti greatly admired, and yet *De Re Aedificatoria* says little about them. When they are mentioned, Alberti makes clear that, though unroofed, gardens are nevertheless like rooms in a house and should be seen as extensions of its central design. The grounds of a great house, he explains, should "also be here and there thrown into those figures that are most commended in the Platforms of Houses, Circles, Semicircles, and the like, surrounded by Laurels, Cedars, Junipers."[14] The same rules that apply to the main building apply to the gardens as well, and they too should be structured according to strict geometrical principles. If anything, since gardens are made of natural objects such as plants, trees, and rocks, and are direct representations of the natural world, it is all the more important that they make visible the hidden geometrical order of nature. Gardens are representations of geometrical order just as buildings are—only more so.[15]

The challenge of turning these general principles into actual gardens was ultimately taken up by members of Lorenzo de Medici's circle in Florence, whose admiration for Alberti knew no bounds. One was Giuliano da Sangallo (1445–1516), scion of a famous clan of architects and engineers, who in the 1480s built Lorenzo a villa and garden in the Medici estate at Poggio a Caiano.[16] The others, needless to say, were Maiano and Fancelli. And so when the ailing Charles VIII took a break from his campaign to recuperate in that Neapolitan villa, we have a pretty good idea of what that "earthly paradise" that stirred his soul looked like: it was, almost certainly, a geometrical garden, laid out in accordance with Alberti's prescribed proportions.

Geometry Goes North

At Poggio Reale, Charles VIII encountered a garden the likes of which he had never seen before. The pleasure gardens of northern Europe, the ones that adorned the estates of the medieval aristocracy, were open space in which trees, meadows, and flowers were scattered freely for the enjoyment of the lords and ladies. But Poggio Reale was as Alberti had decreed—composed of circles, semicircles, squares, and rectangles—and the trees, instead of being scattered freely through the estate, were "planted in Rows exactly even, and answering to one another exactly upon straight lines."[17] Traditional gardens, with their richness and free-form style, showed off the beauty and abundance of the world. "Albertian" gardens like Poggio Reale did that and more: they showed the hidden geometrical design that governs all of creation.

To a visitor such as Charles VIII the experience of looking over an Albertian garden and walking its paths was transformative. For one thing, the garden presented in miniature nearly every aspect of the natural world. André de la Vigne (ca. 1470–1526), the French poet and courtier who accompanied Charles on his Italian adventure, describes the gardens as filled with sweet and beautiful flowers, passages, barriers, slopes, and small rivers.[18] No French garden of the day could compete with Poggio Reale for sheer richness and variety.

But even more strikingly, all this richness and variety was strictly regimented and ordered. Every flower bed, tree, and canal had its carefully assigned place within a general design, and it could not be moved, changed, or tampered with without disrupting the geometrical plan. Each object and its place were determined not by random chance, or even by the whim of a designer, but by the universal laws of geometry. And just as a geometrical circle, rectangle, or triangle was the same for all eternity, so it was for the geometrical garden. It presented a world that was orderly, regimented, perfect, and unchanging. Alberti had argued that "Beauty, Majesty, Gracefulness . . . consist in those particulars which if you alter or take away, the whole would be made homely and disagreeable."[19] And that was precisely the way that the geometrical garden captivated its visitors: it wasn't just beautiful; it was perfect exactly as it was.

To the young French king the perfectly ordered world he found at Poggio Reale must have come as a welcome change. When he entered the gardens in the spring of 1495 he left behind the barely controlled chaos that was the lot of a medieval French monarch even in the most peaceful of times. In the paths of Poggio Reale he discovered a world that was different in every way—a world of order and tranquility, one in which everyone and everything peacefully inhabited their God-given place. Not only that, but any revolt against the harmonious order that infused every aspect of the garden seemed not only foolish, but practically impossible. The garden was designed to perfection according to the universal, unchanging laws of geometry. And no one, not the mightiest nor the wisest, could challenge the laws of geometry.

Charles may not have been the wisest of French monarchs, but for all his limitations he was both competent and energetic when it came to the chief goal of all French kings: establishing effective royal rule over the entire territory that we know today as France. Charles VII, the young king's grandfather, had brought back the monarchy from near extinction at the hands of the English, and his successor, Louis XI, did much to unify the country and expand royal rule. Yet when Louis died in 1483 with his son, Charles VIII, only thirteen years old, it was a signal for great nobles across the land to rise up once more and assert their freedom from their royal master. Even the young king's cousin, Louis d'Orleans, tried to abduct Charles in order to assert his own claims to the regency, if not the crown. It took Charles nearly a decade to pacify his restless kingdom, and when force of arms proved insufficient he took the precaution of marrying his most troublesome subject, Anne, Duchess of Brittany. Even as Queen of France, however, Anne never stopped plotting to regain the independence of her ancient patrimony.

Consequently Charles's rule, much like that of his predecessors and successors, was an incessant struggle to preserve royal authority over mighty subjects, to defend the kingdom from foreign invasion, and to protect the rights of the royal house of Valois against the rival lines of Burgundy and Orleans. His reign was, in effect, an unceasing and unending campaign to establish royal supremacy and royal administration over a land that was perpetually on the verge of revolt. It was a battle to establish order over chaos, a struggle that was both necessary and endless, since final victory was forever beyond reach. For no sooner had one threat been dealt with than another arose, and chaos reigned once more.

At Poggio Reale, Charles learned that it did not have to be this way. The garden was, to be sure, a kingdom in miniature, complete with mountains, rivers, and meadows. But whereas the king's own realm was chaotic and strife-ridden, the miniature kingdom outside the walls of Naples was anything but. When Charles looked at it from the main house, or strolled through its verdant paths, the struggles of his reign were replaced by a peaceful land in which everything had its place, coexisting in an orderly and harmonious whole. The contrast between this and his own kingdom was stark, and Charles could see the cause: France was ruled through a constant and ruthless campaign to suppress disorder; Poggio Reale, in contrast, was governed by the universal, unchanging laws of geometry. In France, a great noble need only wait for a favorable moment to rise up and throw the kingdom into turmoil; but in Poggio Reale, rebellion and unruliness were unthinkable. For who could rise up against irrefutable geometrical order?

Charles VIII was no intellectual, and certainly no theorist of absolutist rule. He was, however, a ruling monarch, and in the quiet Neapolitan garden he seems to have glimpsed something that would aid him in securing his royal prerogatives. What if kingly rule did not have to be a constant campaign of suppression, a barely maintained patina of order over a foaming ocean of chaos? What if his restless kingdom could be as peaceful, orderly, and harmonious as the world-in-miniature at Poggio Reale? What if—like the geometrically arranged paths and trees of an Albertian garden—royal supremacy could become harmonious and natural, an unchallengeable part of the universal order? It was a vision of monarchy unlike anything that had gone before it, and in due time it would become central to the self-presentation and identity of the kings of France. Charles, in all likelihood, could perceive these possibilities only vaguely. And yet, enchanted by the elegant symmetries of Poggio Reale, he resolved to replant this microcosm of harmony and order in the soil of his own kingdom.[20]

And so it was that as the weary French troops made their way back across the Alps in the spring of 1495, they were accompanied by two distinguished Italian gardeners. They were almost all that Charles VIII had left to show for his lightning campaign down the Italian peninsula, and he was determined to put them to good use. One of the two, Fra Giovanni Giocondo of Verona (ca. 1433–1515), was an eminent humanist who had once presented Lorenzo de Medici with an elegant edition of an ancient manuscript. He had apparently worked alongside Fancelli at Poggio Reale in the months leading up to Charles's arrival, but in France, so far as we know, he confined his activities to lecturing and writing on the theory of architecture and gardening. The other, Pacello da Mercogliano (ca. 1455–1534), was less known in his home country. Yet it was he who left his mark on the French landscape by bringing geometrical gardens to the castles of Charles and his successors.[21]

Royal Geometries

Immediately upon arriving in France, Mercogliano received his first assignment: to design and create a new garden in the ancient château of Amboise in the Loire Valley. Situated on a bluff overlooking the Loire River, Amboise is famous today as the burial place of Leonardo da Vinci, brought there in his old age by King Francis I. But to Charles VIII Amboise was special for a different reason: he was born there and—though he did not know it—was fated to die there only a few short years after returning from Italy. Even before embarking on his Italian adventure Charles had begun restoring the old fortified stronghold in the medieval gothic style, familiar to us from the great cathedrals of Notre-Dame, Reims, and Chartres. But Charles was so impressed by the novel classical style he encountered in Italy that he changed course, resulting in the unusual hybrid structure that survives to this day. Mercogliano's mission was to provide this unique château with a garden befitting a great Italian villa.[22]

We do not know the precise outlines of Mercogliano's creation because unlike buildings, which can sometimes remain unchanged for centuries, gardens are in a perpetual state of flux. Without constant human care and supervision they would not survive from one season to the next, not to mention year to year or one century to another. Consequently no garden we know today, not even old and established ones like Tivoli or Versailles, is truly the same as the historic garden that thrived in the same place centuries ago. To reconstruct ancient gardens we are dependent on sources from their time, or at least as close to it as possible. In Mercogliano's case this means relying on the French architect Jacques Androuet du Cerceau (1510–1584), who in the 1570s published two great volumes of engravings depicting the great palaces and châteaus of France.[23]

Judging from du Cerceau's engravings, the Italian master did his best to accede to the monarch's wishes. A rectangular open terrace was set aside for the garden within the walls of Amboise, on the northeastern side of the castle, overlooking the river. Mercogliano divided this area into ten evenly spaced rectangles, surrounded by railings and arranged in two parallel rows of five each. Eight were of roughly equal size, with two smaller ones capping the western end. A wooden pavilion sheltering a fountain separated the second and third rectangles in the northern row, along the castle walls. Every one of these larger rectangles was in turn divided into four equal small rectangles, which were each planted in carefully designed geometrical patterns—circles, squares, ovals, and so on. We can only guess what Mercogliano had planted there, but most likely it was flowers and shrubs rather than the more practical vegetables. A pleasure garden, after all, marked its owner as one who could afford to devote precious space and resources to his own amusement and that of his companions, rather than to daily sustenance. It was, by definition, the opposite of useful.

If Charles VIII wanted a geometrical garden at Amboise, he certainly got one. The shape of the terrace and its divisions, the symmetrical arrangement, the rectangular parterres, and the patterns within them are all arranged in a strict geometrical order. Yet there was no denying that Mercogliano's garden was no Poggio Reale. The Neapolitan garden, after all, was home to trees, bushes, meadows, flowers, and streams much like those that could be seen in the surrounding countryside. Its magical power was derived from the fact that all these common features were arranged in regular patterns, thereby making visible the hidden geometries that pervade all.[24]

Mercogliano's garden at Amboise, in contrast, was enclosed within the château and conformed to the military architecture of the castle, not the natural geometries of the surrounding countryside. Perched on a perfectly flat platform, artificially raised above the river and town, and surrounded by fortified walls, it was entirely separate from the landscape of the Loire Valley. So much so that even though the garden was perched right above the river, a lord or lady walking its paths would see nothing but imposing stone walls, while a peasant looking up from the riverbank would never know that the garden was even there.

Furthermore, when it came to the garden itself, Mercogliano's design did little to suggest any correspondence between the castle garden and the surrounding countryside. The garden at Amboise, so far as we know, had no trees, and the only suggestion of streams and rivers was a single enclosed fountain. Nothing remained of the "passages, barriers,

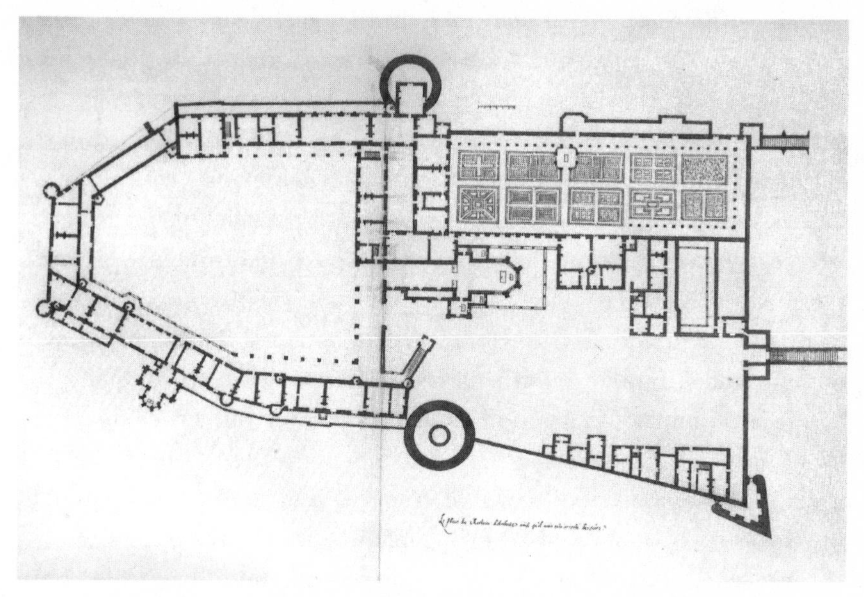

FIGURE 12: **The Château and Gardens of Amboise in the 1570s according to du Cerceau**

slopes, and small rivers" that had so enchanted André de la Vigne at Poggio Reale. A visitor to the garden at Amboise might wonder at the elegant geometrical patterns of the parterres but would have no inkling of the rich, harmonious universe that Charles had glimpsed at Poggio Reale. If geometrical gardens were to serve as public emblems of an orderly and peaceful kingdom, then the garden of Amboise was useless.

Kings and Gardens

Whether Charles VIII was pleased with what Mercogliano created for him at Amboise will forever remain a mystery. In April of 1498, with work on the garden in full swing, the king struck his head on a door lintel in the castle while rushing to view a game of tennis. Lesser souls might have taken to bed, but Charles was an ardent sportsman and, undeterred, he sat and cheered through the entire match. On his way back to his apartments he suddenly collapsed, and he died in his chambers a few hours later. He never laid eyes on the finished garden.

Since Charles died without an heir, the crown passed to his cousin Louis d'Orleans, the same man who had tried to abduct the young king fifteen years before. Louis XII (1462–1515), as he was now known, was thirty-six years old on his accession, and he brought to the throne a degree of experience and worldliness that Charles had conspicuously lacked. Intent on consolidating his kingdom, Louis had his marriage to Princess Jeanne of France annulled by the pope and immediately married his predecessor's widow, Anne, mistress of the troublesome Duchy of Brittany. For the next seventeen years he ruled his kingdom with wisdom and frugality, reforming the royal accounting, improving the administration of justice, and, most popular of all, lowering taxes. Challenged for his parsimony by lords used to the lavish life of Charles's court, he responded that he would "much prefer to make fops laugh at my miserliness than to make the people weep over my generosity."[25]

Yet when it came to his foreign ambitions, and in particular with regard to Italy, Louis followed precisely in Charles's footsteps. In 1501, taking advantage of the endemic squabbles among the Italian powers, he too led his army across the Alps and swept down the peninsula, occupying first Milan and then Naples. Once again, things did not turn out well: like his predecessor, Louis was soon defeated by King Ferdinand of Aragon, and by the end of his reign the French foothold in Italy had disappeared entirely. For the second time in twenty years a seemingly invincible king of France found himself foiled by the inscrutable politics of the Italian peninsula.[26]

Louis did not share his predecessor's romantic infatuation with Italian art and luxuries, but he did recognize the importance of what Charles had created at Amboise. In 1499 he purchased a rectangular tract of land, approximately two hundred meters by seventy-five meters, outside the royal château of Blois in the Loire Valley, and instructed Mercogliano to turn it into another geometrical garden. Just as Amboise was special to Charles, Blois was special to Louis, who had been born in the château and spent his childhood there. Geometrical gardens, it seems, were an honor bestowed by the monarch only on the most cherished of his many dwellings.

The garden at Blois, as seen in du Cerceau's engraving, includes a small lower garden and a trapezoid-shaped upper garden, but it is the middle level, built on the land purchased in 1499, that is Mercogliano's creation. It is this garden that is unquestionably the child of Amboise: both gardens were elegantly divided into rectangular compartments in two parallel rows, with a central axis down the middle. Both were surrounded by a wall and insulated from the surrounding landscape, creating inward-looking worlds meant exclusively for the enjoyment of the residents of the château. Yet the garden at Blois also included some important innovations. For one thing, it was located outside the main castle rather than on a high terrace inside the fortifications, and was set at different levels, following the contours of the land. This made the garden more visible and more accessible from the surrounding hills, and better integrated into the landcape. For another, Dom Antonio Beatis, who visited in 1517, reports that he saw "many lemon and orange trees in wooden tubs," which were kept in a wooden shed during the winter and brought out in the spring. The flowers and shrubs of the garden of Blois were supplemented with exotic trees, suggesting a more sustained effort to re-create the rich variety of the natural world.[27]

By creating at Blois a garden that is more open and more diverse, Mercogliano managed to correct some of the deficiencies of Amboise. The modifications, to be sure, were relatively minor, yet the design trends manifested at Blois were a sign of things to come: over the following decades French royal gardens would grow ever larger, richer, and more diverse, and increasingly open to the surrounding countryside. Instead of secretive enclosures where monarchs and their servants took refuge, they would become representations of a harmonious geometrical order that governs the kingdom and all within it.

The Brilliant King

When Louis too died without a son in 1515, he was succeeded by François d'Angoulême, his dashing son-in-law (and second cousin). Francis I (1494–1547), as he now became, was the epitome of French chivalry: handsome, strong, and energetic, an enthusiastic participant in jousting tournaments and an ardent sportsman, he was known for amusing himself by fighting wild boars single-handedly.[28] He was also, however, the greatest patron of arts and letters France had ever known. A passionate admirer of Italian art and culture, he invited the aging Leonardo da Vinci to make his home in the grand house of Clos de Lucé, adjoining Amboise. Showing impeccable taste, the king purchased the *Mona Lisa* from Leonardo's assistant and hung it in his palace of Fontainebleau. Other famous dependents included the Florentine painter Andrea del Sarto (1486–1530) and Benvenuto Cellini (1500–1571), the tempestuous Florentine goldsmith and sculptor. Having been educated by leading humanists, Francis endowed royal lectureships in the classical languages, which would become the foundation of the Collège Royal, today's Collège de France. To fill the shelves of his royal library he sent his agents to Italy to purchase the latest and finest humanist editions. He then opened the library doors to scholars and invited them to make use of its vast and growing collection.

Unfortunately for the people of Italy, Francis's fascination with their land—like that of his predecessors—was not limited to its art and culture. Only a few months after his accession in 1515 he led his army across the Alps, crushed his enemies in the battle of Marignano, and occupied Milan. It seemed like the beginning of a glorious reign, and it might have turned out that way were it not for an event that took place four years later: in 1519, much to Francis's dismay, the Habsburg Prince Charles of Austria was elected Holy Roman Emperor.

At the time of his election Charles V was already archduke of Austria, lord of the Netherlands, duke of Burgundy, and, most important, king of Spain. Suddenly Francis found his kingdom surrounded by a mighty empire that included not only Germany to the east and Spain to the west but also territories stretching from the spice islands of Southeast Asia to Peru in the New World. Despite his pugnaciousness and martial spirit, the French king could not match the resources of Charles's globe-spanning possessions, and he spent the remainder of his reign locked in a near-continuous defensive war. At the lowest point of his fortunes, in 1525, Francis's army was overwhelmed by Charles's Spaniards in the battle of Pavia, and he himself became the emperor's prisoner. He regained his freedom by marrying his captor's sister and providing his sons as hostages, and then proceeded to resume the war as if nothing had happened.

Or so it seemed. For the king who returned from the Spanish captivity was a changed man, intent on reforming the structure of the monarchy. Before his captivity Francis had been a restless soul, a man without a home who crisscrossed his territories incessantly, north to south and east to west. He moved constantly from one château to the next, from town to town and from one noble estate to another, accompanied by a royal court that numbered as many as 10,000 men and women.[29] One town that he seldom visited, however, was Paris, the ancient capital and—with a population of around 200,000—by far the largest city in the realm. This is not as strange as it sounds: the Renaissance kings of France relied for their rule on the provincial nobility and were consequently required to make their presence felt in the lands of their mighty subjects. A king that neglected to both show his respect to his great nobles and overawe them with his personal presence would soon have to deal with an armed rebellion. The people of Paris, in contrast, were neither an important asset nor a dangerous threat, and could safely be ignored.[30]

Yet Francis's defeat and imprisonment had convinced him that things could no longer continue as they were. The great nobles, cowed by the threat of foreign invasion, united behind the king. But the Parlement of Paris, the highest law court in the land and the one charged with ratifying royal proclamations, had taken advantage of the king's absence to challenge his mother, who was acting as regent. Meanwhile the Reformation, which had started in Germany, had now reached France and was spreading swiftly among Francis's subjects without regard to lineage, class, region, or gender. Lutheranism, and later Calvinism, were infecting both nobles and commoners and stirring up religious divisions and long-dormant passions.

Faced with the threats of social subversion in Paris, a new faith that seemed to be sweeping all before it, and constant pressure from his nemesis, Charles V, Francis realized that he must act to preserve his kingdom. In April of 1527 he appeared in Paris, arrested the leaders of the rebellious Parlement, and overrode its decisions by royal decree. The following March he made his intentions clear: "Dearest and well-beloved," he wrote to the Paris town council, "it is our intention henceforth to stay most of the time in our good city of Paris and its neighborhood."[31]

Old habits die hard, and in the end, even after his brave proclamation, the king continued to travel with his court from one end of the realm to the other. Yet the focus of his attention had clearly shifted, and with it the direction that the French monarchy would take over the next two centuries.

The kings who favored the magnificent châteaus of the Loire Valley were itinerant monarchs. Charles VIII, Louis XII, and the young Francis knew that power in their kingdom was held by the rural nobility and spread all across the land. And so they did their best to be present everywhere. But by promising to reside in Paris, Francis was offering a different vision of monarchy, one based not on omnipresence but on a permanent seat of royal power. With the king and his court in regular residence, Paris would become a true capital, the stationary center of administration on which all lines of power converge. All power and legitimacy would now be invested with the king and his court, not with the landed aristocracy of the provinces. Now it would be up to the nobles to make their way to the royal court, the fount of all power, honors, and legitimacy.

All this, to be sure, was a distant dream in the age of Francis I, as even the most far-seeing observer could not yet imagine the centralized monarchy that would emerge in France toward the end of the following century. And yet Francis had set a new course for the monarchy, a course that his successors would continue to follow from generation to generation. Sometimes hesitantly, at other times confidently, the kings of France would slowly create the concentric kingdom that Francis had but hinted at. And no emblem of kingship and state represented this ideal or propagated it more powerfully and effectively than the French royal gardens.

We know little of Francis's taste in gardens: Chambord, the dreamlike château he built in the Loire Valley, was set in the midst of a forest, and if the Château de Madrid near Paris had a garden, we know nothing of it. But Francis's favorite château in his later years, the one in which he housed his library and where he spent the most time, was Fontainebleau, and it was there that he created the most elaborate royal garden of his reign. Situated near the town of Melun, forty miles south of Paris, Fontainebleau was one of the oldest royal keeps, and a popular hunting lodge for several generations of kings. When Francis vowed to make his home primarily in Paris he began a vast building project on the grounds of Fontainebleau in which gardens played a central role. By the time of the king's death in 1547 the old castle had been transformed into a full-fledged Renaissance palace, and it was almost completely surrounded by gardens.[32]

This in itself was significant. The garden at Amboise was on a terrace within the castle, the one at Blois to the side of the château at an oblique angle. In neither case did the château itself have any bearing on the design of the garden, and it was hardly noticeable from within it. At Fontainebleau, in contrast, the main courtyards of the château open directly onto the lake and gardens, making the landscaping a direct continuation of the palace architecture and the palace the main focus of the garden. It is the first time that a royal garden places the château, and implicitly the king, at the center of its design. The implication, that the natural order is centered on the monarch, is only gently hinted at in Francis's palace. Yet it would be an increasingly dominant theme in future royal gardens, and would reach its climax in Versailles.

The first garden at Fontainebleau, created around 1530, was a relatively small square to the north of the oval main palace. Like the ones at Amboise and Blois, it was divided into four equal square parterres, each elaborately cultivated and patterned into geometrical figures. Rectangles, regular triangles, concentric circles, and equally spaced radii emanating from the center are all present in du Cerceau's engraving, suggesting a veritable geometrical kingdom. The main gardens of Fontainebleau were, however, to the south of the palace and were dominated by the central feature of the estate: a large trapezoid-shaped lake with perfectly straight sides, opening up to a palace courtyard called Cour de la Fontaine. It was the first time that a large body of water was

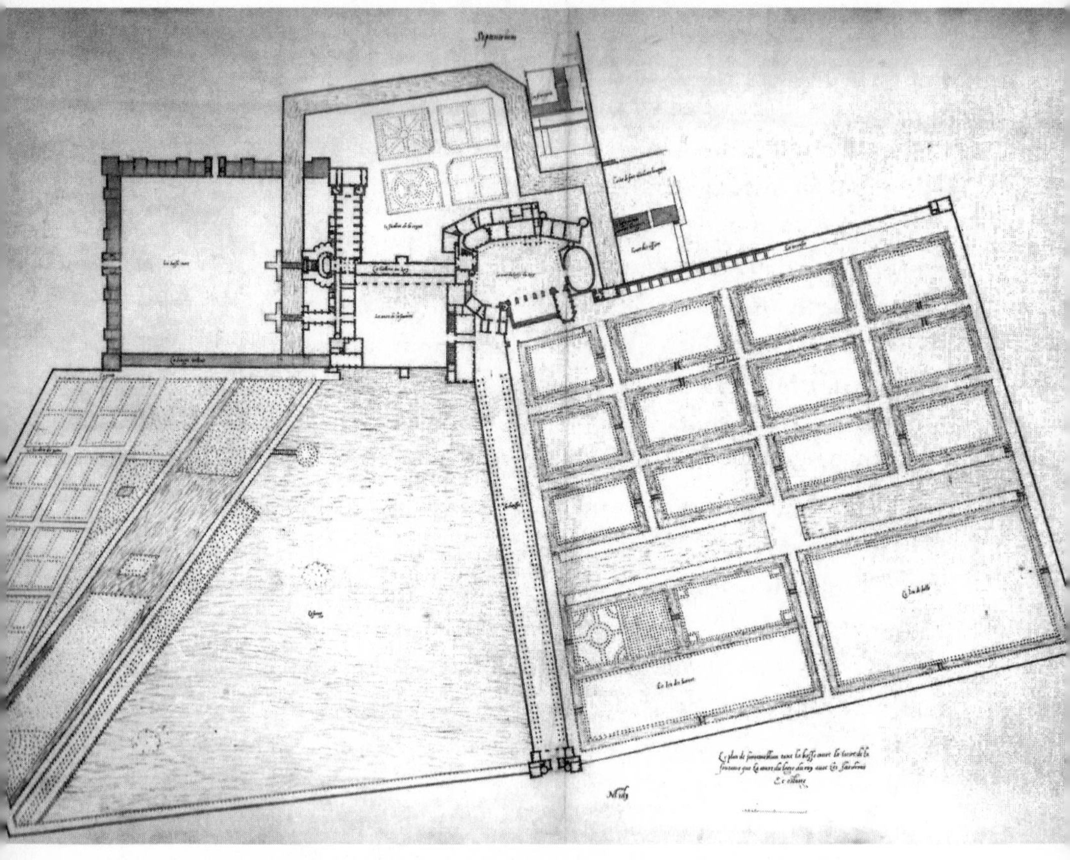

FIGURE 13: The Château and Gardens of Fontainebleau in the 1570s, according to du Cerceau

included in the landscaping of a royal château, and its contours determined the outlines of the surrounding gardens.

On the western side of the lake Francis purchased a trapezoidal orchard and transformed it into a garden. Between 1535 and 1538 the ground was leveled, canals dug, and the oddly shaped area divided into regular rectangular plots, supplemented by triangles to accommodate the garden's unusual contours. It became known as the Jardin des Pins, after the many pine trees within its borders. Along the eastern side of the lake, beyond the main causeway leading to the grand entrance to the palace, Francis acquired a rectangular tract of land and transformed it into a vast square Grand Jardin. Many times the size of the northern King's Garden, the Grand Jardin was divided into two unequal rectangles by an east–west canal. The northern segment, following the precedent set in the castles of the Loire Valley, was divided into evenly spaced rectangular parterres, twelve in number, whereas the southern segment was divided into only two large rectangles.

The southern gardens of Fontainebleau were, to be sure, geometrical in nature, but of a rather different kind than the ones we encountered in the Loire Valley. At Amboise and Blois, as well as in the King's Garden at Fontainebleau, the gardens are precisely measured rectangles, divided into rows of identical square parterres. Each of these is then patterned into elaborate figures, thereby creating what is effectively a self-contained Euclidean haven. Insulated from the natural world all around them, these spaces were oases of geometrical reason, order, and precision in an ocean of chaos and bustle. They were, in effect, quiet corners of Euclid's universe or Plato's realm of the forms, fully separated from our daily reality.

But the Pine Garden and the Great Garden at Fontainebleau were different: with their wide expanses of meadows they were not as carefully manicured as the gardens of the Loire Valley, and with their pine trees, willows, and vines they resembled not so much a pristine geometrical universe as a more orderly version of the hills and woods that surrounded them. Then as now, the effect is reinforced by the low, and in places nonexistent, walls surrounding the gardens, which make them easily visible to anyone passing by the estate and make the countryside a constant background to anyone walking or riding within the gardens. Finally, the large lake separating the two gardens echoes the water-rich landscape of the Île-de-France, including the nearby Seine River and its tributaries. In contrast to Amboise and Blois, the interplay between the artificial layout of the palace grounds and the surrounding natural landscape is palpable.

At Fontainebleau the enclosed royal garden that Mercogliano had created at Amboise had broken through the castle walls and into the natural world beyond. Like the outside world, the gardens contained trees, flowers, meadows, hedges, and—most impressively—a large lake stocked with fish. They represented nature in all its glorious variety, as seen beyond the garden's walls. And yet nature, as one encounters it in these gardens, is not chaotic, but ordered, even regimented, arranged in squares, rectangles, triangles, and circles. The seemingly boundless richness and variety of nature is here governed by a strict and rational geometrical order. This is true not only of the gardens but of the open country as well, since at Fontainebleau the borders between the two were permeable. The geometrical order of the gardens, this implied, might not be as visible in the countryside as it is on the palace grounds. And yet it extends far beyond the garden walls and, implicitly, to all of nature. For the world, as Alberti had proclaimed a century before the first trees were planted around the palace of Fontainebleau, is geometrical to the core.

The continuity between garden and open fields was central to the design of Fontainebleau, but in making this point the gardens' creators compromised on their chief principle: geometricism. The landscaping at Fontainebleau was certainly geometrical, but—apart from the relatively small King's Garden—only in its outlines. In their overall shape the gardens were square and trapezoidal, the parterres rectangular and occasionally triangular. But within each of these parterres the actual arrangement of the flowers, shrubs, and trees was apparently free-form, harking back not to Alberti's aesthetics but to the medieval flowery mead. In their openness to the world, their vast scale, and their richness and variety, the gardens of Fontainebleau represented a vital step in the development of French gardens from Amboise to Versailles. But in presenting the deep geometrical architecture of nature they were a compromise, and perhaps even a step backward. In future years royal gardeners would work to reconcile an openness to the natural world with an insistence on strict geometrical patterns.

The Death of Chivalry

Fontainebleau was surely the most beloved of Francis I's many dwellings, as well as the home of his chief mistress, Anne, Duchess of Étampes. But he did not end his days there: the fever caught him on a journey, during a stay in Rambouillet, another of his Île-de-France residences, some thirty miles from Paris. A temporary reprieve allowed him to go hunting one last time in the forests of that ancient château, but upon his return his condition worsened, and he died on the morning of March 31, 1547. Francis was fifty-two years old, and still trying to fulfill his promise to settle down in his capital, or at least remain in its vicinity. And yet, as his last journey indicates, he remained to the end, like all his predecessors, a nomadic monarch.

Francis's successor, King Henry II (1519–1559), was not raised to be a king. As the old king's second son, he was thrust unexpectedly into the line of succession when his older brother, Francis, died in 1536. Yet when he succeeded to the throne in 1547 he appeared to be ready. Strong and decisive, he kept the squabbling nobles of his realm under control and continued the policies of his late father. At home this meant keeping in check, as much as possible, the spread of Calvin's heresies emanating from Geneva and preserving France as an orthodox Catholic kingdom. Abroad, this translated to continuing the long feud with the emperor Charles V, a policy that once again involved repeated incursions into Italy.[33]

Like his father, Henry was an enthusiastic jouster and hunter, styling himself as the epitome of chivalry, but he inherited none of Francis's intellectual ambitions or love of the arts. His reign was neither glorious nor glamorous, but he succeeded at what was the chief requirement of a French monarch of his day: keeping his kingdom from disintegrating. It wasn't easy. Quite apart from the constant war against the Habsburgs, Henry had to deal with the growing enmity between the noble Guise and Montmorency families, as well as with the increasingly independent Bourbons, blood relations of the ruling Valois. Each of these clans controlled vast territories within the kingdom, and abundant revenues, and could lay claim to the loyalty of officials not only in their domains but also within the royal administration. Add to that the seemingly unstoppable spread of the new Calvinist faith at all levels of French society, creating what could only be viewed by the king as a cancerous tumor growing within the kingdom's body politic. Efforts to suppress the new religion mostly backfired, increasing not only the zeal of the new converts but also their solidarity across regions and class lines. Mysteriously, they came to be known as Huguenots.[34]

For twelve years Henry managed to impose a semblance of political order on the chaos that was bubbling just beneath the surface of his unruly kingdom. That is, until June 30, 1559, when he took part in a jousting tournament held in the rue Saint-Antoine in Paris. The king was a glorious sight: covered from head to toe in brilliant armor and wearing the black-and-white colors of his mistress, Diane of Poitiers, Henry rode his horse Malheureux ("Unlucky") to the top of the lists and, lance held high, challenged all comers. Such bravado was typical of Henry, and seemed perfectly justified. Charging down, he dispatched first the Duke of Nemours, then the Duke of Guise, and then faced a tougher challenge: Count Gabriel de Montgomery, the twenty-nine-year-old captain of the king's Scots Guard. A first clash of the two horsemen proved indecisive, and at that point the king could have retired from the day's events with honor. But Henry would have none of it: "I want my revenge," he shouted, and charged down the lists at Montgomery without even waiting for the marshals' trumpet call. The clash shattered both lances and sent riders and horses sprawling on the ground. The young count scrambled to his feet and quickly mounted his horse, but the king could not follow. He stumbled and then collapsed, bleeding profusely through his visor. From his eye, head, and neck protruded long wooden splinters, the remnants of Montgomery's lance.

Henry did not die that day. For more than a week he clung to life, long enough to forgive Montgomery, who had begged to be punished, and long enough even to dictate two letters. Ultimately, however, even the ministrations of the best physicians in Europe could not save him. His condition took a turn for the worse and he died on July 10, 1559, ten days after the clash of arms. It was a fitting end for a man who prided himself more on his manly prowess than on his intellectual gifts, and for a monarch who relied on force and stamina to keep his restless kingdom from splitting apart.

Just how desperate and prodigious Henry's efforts had been soon became clear when, with the king's steadying hand suddenly removed, Frenchmen found themselves at one another's throats.[35]

4.

THE RETURN OF THE KINGS

The Monarchy in Crisis

When Henry II was killed at the point of a spear in the summer of 1559, it was not just he who died that day. That tragic clash of arms was also the final gasp of an ideal that Henry had inherited from his forebears going back centuries: the vision of the king as the paragon of chivalry, the knight in shining armor who was the most brilliant, most powerful, and most courageous noble in the land. Henry, challenging all comers and charging down the lists on his aptly named horse, brilliant in the colors of his lady love, was the very embodiment of this ideal. For Henry, just as for Charles, Louis, and Francis before him, kings were the leaders of the landed nobility. They not only relied on the aristocracy to support and enforce their rule, but also shared the nobles' values and chivalric ethos.

But when Henry, the last of the knight-kings, was unhorsed in the rue Saint-Antoine, the monarchy that he had known and helped sustain was struck down with him. After Henry, a king who was simply the leader of the nobility could no longer contain the violent tensions between regional, class-based, and religious factions in the body politic, and the land descended into a bloody civil war that would last to the end of the century. When, following nearly four decades of fratricidal carnage, the French monarchy reemerged victorious, it was a very different institution: instead of leader of the nobility, the king would become the focal point of the entire nation, ruling the land directly from his permanent seat in Paris. It was a slow and uneven transition, but when in 1661 Louis XIV announced that he would rule his kingdom directly, without a chief minister, it was complete. The centralized monarchy, which Francis I had dimly envisioned when he promised to move his residence to the Paris region, had become a reality.[1]

In 1559, however, all this lay in the distant future, as France was flung into one of the deepest crises in its history. Sensing a power vacuum at the top, the previously secretive Huguenots came into the open to challenge their Catholic brethren. Spurred along by Calvin from his seat in Geneva, just across the Swiss border, they were determined to take advantage of the opportunity to convert both court and country. Alarmed Catholics, lacking a strong king to protect their cause, banded together into mobs and hounded the Huguenots from towns and territories where they were a minority. The result was a vicious cycle of expulsions, massacres, rapes, and looting on both sides, which continued unabated for thirty-five years. The St. Bartholomew's Day Massacre of August 1572, when rampaging mobs butchered thousands of Huguenots in Paris and across France, was only the worst and most famous of a relentless stream of atrocities.[2]

Both sides, as was the habit of the time, looked to the great nobles for leadership of their cause. The Guise clan, dominant at court since the days of Francis I, took up the defense of the old faith, whereas their rivals, Admiral Coligny and the prince of Condé, converted to Protestantism and became de facto leaders of the Huguenots. Meanwhile, defying both nobles and kings, city artisans and lower clergy banded together in the fiercely militant Catholic League, which brooked no compromise with those they considered heretics. Clan rivalries and class enmities had racked France for centuries, particularly under weak monarchs, but to these was now added a new cause: religious fervor. And as religious truth is absolute and unyielding, so the prospect of reaching a compromise peace between the rival factions was ever more remote.[3]

As the kingdom faltered, the French monarchy staggered along. The first of Henry II's sons to ascend to the throne was Francis II, who although a nominal adult was only fifteen years old on his accession, and he was easily manipulated by his leading courtiers of the house of Guise. When he died after only eighteen months on the throne he was succeeded by his younger brother Charles, who was only ten. Sidelined in the early years by his mother, Catherine de Medici, Charles IX lived to reign for fourteen years, but proved as ineffective a ruler as his brother. Desperate to keep his kingdom whole but distrusted by both sides in the war, he veered back and forth in his policies, alternating between pragmatic accommodation of the Huguenots, as was championed by his mother, and their outright extermination, advocated by the Guises. It was under his watch that a royal celebration, to which the leading Huguenot princes were invited as a sign of reconciliation, turned into the bloodbath of St. Bartholomew's Day.

Charles died without an heir and was succeeded by his brother Henry (1551–1589), the third and most promising of Catherine de Medici's children to ascend to the throne. Henry III was intelligent, learned, and reportedly a born orator, yet he was no more able to maintain peace in the kingdom than his brothers had been. Although personally inclined to moderation, he could not shake the uncompromising Guise courtiers' hold over royal policies. The result was that more-moderate leaders, such as the king's brother the Duke of Alençon and Henry, King of Navarre, withdrew from the court and kept much of southern France outside the reach of royal power. When the king signed a peace treaty with the Huguenot rebels, he brought upon himself the wrath of the people of Paris, who joined the Catholic League to defend their

Roman faith. Their rage at the pacific-minded Henry mixed with panic at the knowledge that if the king remained childless, he would be succeeded by the Protestant Henry of Navarre.

True to his nature Henry tried to appease the Catholic faithful, but in 1588 the Parisians rose up and expelled him from the city while simultaneously welcoming their champion, the Duke of Guise. For the king, this was the turning point: he lured the duke and his brother, Cardinal Guise, to Blois, where he was staying, and promptly had them murdered. He then joined forces with Henry of Navarre to lay siege to the rebellious capital and crush the traitorous League. Whether the two Henrys would have succeeded will never be known. Henry III was assassinated by a Dominican friar on August 2, 1589, the night before he was to lead an assault on the city.

Thus came to an end the disastrous reigns of the three children of Henry II and Catherine de Medici. In 1559 Henry II bequeathed to his successor a peaceful kingdom in which the authority of the monarch was everywhere acknowledged. Thirty years later Henry III left behind a kingdom in shreds: the capital and many towns were under the sway of the Catholic League and in open revolt against the king, while much of the south was under the sway of Protestant heretics. It was the kingdom's deepest crisis since the darkest days of the Hundred Years' War.

Yet throughout this time, even as private armies and religious militias ransacked the land and fought pitched battles, as Catholic and Protestant mobs assaulted and butchered one another, as the authority, prestige, and finances of the monarchy collapsed, the French monarchs persisted with the activity that had defined them since the days of Charles VIII: building palaces and—more specifically—geometrical gardens. The person most responsible for this was not, in fact, one of the kings, but their mother, widow of Henry II, the formidable Catherine de Medici (1519–1589).

The Mother of Kings

As was usual among the great clans of Europe, Catherine's marriage to Henry was a matter of politics, not love. It came about when Francis I,

Henry's father, was seeking an alliance with Pope Clement VII, previously known as Giulio de Medici. Yet despite this far-from-romantic beginning, the fourteen-year-old bride adored her young prince, even as he spurned her and took up with a series of mistresses. For the last twenty years of their marriage she had to share him with Diane of Poitiers, a widow nineteen years his senior who became not only his mistress but his close confidante. Even worse, when Henry in 1536 unexpectedly became heir to the throne, Catherine's failure to produce a son put the royal succession at risk. She would likely have been cast aside were it not for the protection of old king Francis, who was fond of his spirited daughter-in-law. Only in 1544, more than a decade after her marriage, did Catherine give birth to her first child, the future Francis II. In the following years she bore four more sons and five daughters, seemingly securing the Valois line.[4]

Though her position as the queen mother of the royal family was now assured, Catherine had little influence on policy making during her husband's twelve-year reign. She suffered silently as Henry spent a good part of every day with Diane, consulted her on all things, and praised her charms to his courtiers. Yet when Henry was mortally wounded, she had her revenge: she banished the royal mistress from the dying king's bedside and then from the court, and quickly expelled her from Chenonceau, the Loire Valley château Henry had gifted her. Then, with her young sons succeeding one another on the throne, she plunged head-on into the murky waters of sixteenth-century French politics.

It is difficult to assess exactly how much impact Catherine had on royal policies during the thirty-year reign of her three sons. Other than the three years when Charles IX was a minor and she served as regent, she had no official role in the government and could influence events only to the extent that she could prevail upon her sons to follow her advice. Yet her prestige was enormous, not only as the mother of successive kings, but also as a link to the happier age of Francis I and Henry II. For the most part she sided with the *politiques*, the court faction who sought to reach a pragmatic settlement with the Huguenots, and against the uncompromising stance of the Guise clan. But increasingly, as her overtures to the Huguenots failed to bring about the desired peace, she endorsed harder-line policies, and she could be ruthless when she thought it necessary. For thirty years, as kings, councilors, and courtiers came and went, as some rose and fell and not a few were assassinated, Catherine was a constant, reassuring presence. More than just a player in the bloody political games of the age, she was the living embodiment of the continuity of the monarchy, a promise that beneath the chaos the heartbeat of the kingdom persisted, waiting to reemerge.

And indeed, no one was more committed to preserving the flickering flame of royal authority than Catherine. As she maneuvered deftly between the opposing factions, veering back and forth between accommodation and oppression, she also pursued a steady cultural policy that never changed course and never wavered. If the monarchy was to recover, she knew, and the kings to reclaim their authority, it was critically important to keep up a glorious display of royal magnificence for all to see. This was not a matter of keeping up an empty façade of royal power at a time when it was more fiction than fact. It was, rather, a keen awareness that public spectacle was inseparable from royal power, that extravagant displays of wealth and luxury were not wasteful, but instead served to establish the legitimacy of a king and his place as the natural and inevitable ruler. For who but a king would be capable of such display, or worthy of such magnificence?

And so it was that Catherine pursued an unwavering policy of royal magnificence. While this was important at any time, it was particularly critical at a time when the monarch's authority was being challenged from all sides. The king might be at war with his subjects and his power might not extend far beyond his court, but as long as he has grand palaces, fabulous entertainments, and exquisite artworks, he is still king. And so, repeatedly over three decades, the queen mother sponsored grand public entertainments involving mythical nymphs and sirens, provocatively obscene satyrs, and courageous knights riding to the rescue of virginal shepherdesses. She commissioned the great Michelangelo to carve a statue of her late husband, Henry II, on horseback, though the old master excused himself on account of his advanced age. She built, though never completed, a chapel of the Valois at the Abbey of St. Denis, traditional resting place of French royalty. A monumental tomb of Henry II, with Catherine by his side, was to be its centerpiece.[5]

Catherine was the great-granddaughter of Lorenzo the Magnificent, and so it is perhaps not surprising that her most ambitious projects were architectural. After taking possession of Chenonceau, she renovated the castle and embellished the gardens with waterfalls, an aviary for exotic birds, and a menagerie of rare animals. She also rebuilt and renovated the castles at Montceaux-en-Brie, near the town of Meaux, and at Saint-Maur, near Vincennes, southeast of Paris. But her grandest and most enduring legacy was the building of a sparkling new palace with a giant geometrical garden just outside the walls of Paris. Massive, majestic, and located in the royal capital, it did more than any of Catherine's projects to enhance the glory of the kings of France. Built on land formerly used for the manufacture of bricks and tiles (*tuile*), it was called, appropriately, the Tuileries.[6]

The Tile Factory

Like so much of the tragic history of France in the age of Catherine de Medici, the Tuileries have their origins in the violent death of Henry II. Deeply shaken, the queen refused to stay another day in the Tournelles, the royal residence in Paris where the king breathed his last. She quickly packed up her household and children and moved to the Louvre, on the northern bank of the Seine, just inside Paris. The Louvre of the sixteenth century, however, was not the kind of place to please a queen conscious of the prerogatives of her rank. Although refurbished in the final years of Francis I, it was still a combined military fort and royal prison, and Catherine soon grew tired of its manifest lack of grandeur. Looking for a more worthy seat of royal authority she happened on the Tuileries, along the river to the west of the Louvre, but just outside the city walls. Purchased by Francis I as far back as 1519, the estate was largely undeveloped, and it seemed to Catherine like the perfect locale. She plucked her husband's favorite architect, Philibert de l'Orme, from forced retirement and commissioned him to design and build her dream royal palace.

In 1564 de l'Orme set to work, and for the next eight years the palace and garden grew at a rapid pace. Du Cerceau's engraving of the palace from 1576 shows it to be a grand structure built around three rectangular courtyards, with a façade looking westward, away from the city. The image, though, is likely more concept than reality, because in fact Catherine's Tuileries were never completed. According to legend, a fortune-teller warned the queen to avoid Saint-Germain if she valued her life, causing her to abandon the project because the Tuileries were in the parish of Saint-Germain. But since Catherine did, in the following years, stage numerous celebrations in the gardens, seemingly unconcerned about dangers to her life, it is more likely that the cause was lack of funds. Whatever the reason, in 1572 the palace stood only half-finished, and was to remain so for decades.[7]

But while the palace remained only partially habitable, the gardens bloomed, and by the 1570s they were already viewed as the paradigm of a royal garden. According to du Cerceau's engraving, the gardens were roughly rectangular, with the shorter side paralleling the façade of the palace and the longer side stretching westward along the river.[8] The western border was at an oblique angle to the southern side along the river, making the precise shape a trapezoid. As was the case in royal gardens from Amboise to Fontainebleau, the area was carved into squares and rectangles, but on a larger scale than anything seen before: the garden at Amboise had ten squares, the Grand Jardin at Fontainebleau about seventeen. The Tuileries had thirty-eight.[9] The variety among them was also unprecedented: "[T]he arrangement of trees and plants is admirable," reported the Venetian ambassador. "Not only do we find mazes, bosquets, streams, and fountains, but even the seasons of the year and the signs of the zodiac are represented. That is truly a marvel."[10]

Even more than at Fontainebleau, a conscious effort was made to replicate the richness and variety of the kingdom within the gardens: In 1566, 500 fruit trees were planted, and irises (crocuses) were added in 1568. Four years later 100 pear trees, 50 almonds, 150 cherry trees, 150 plums, and 425 elms and limes were added, and the main passageways in the garden were lined with sycamores, elms, and fir trees.

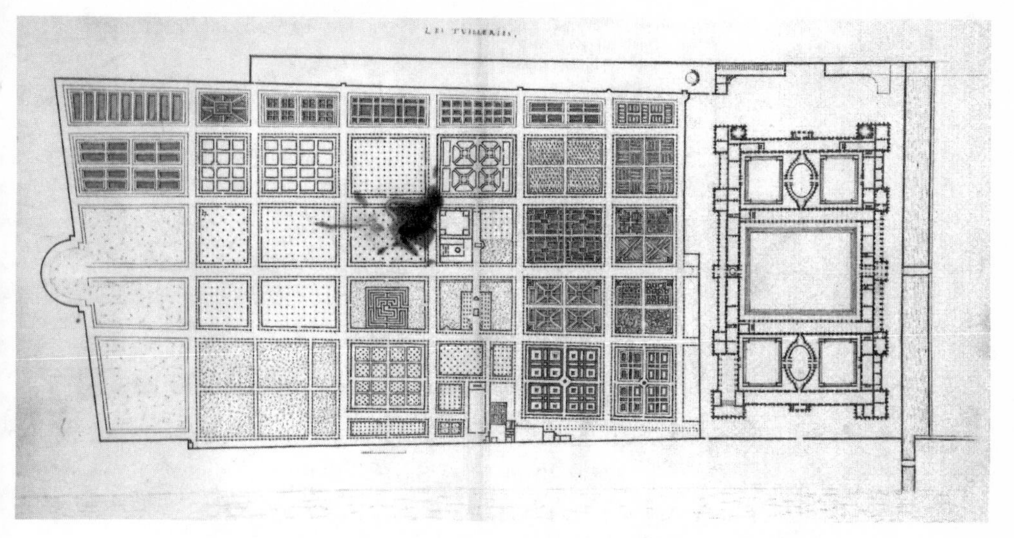

FIGURE 14: **The Tuileries in the 1570s, according to du Cerceau**

Catherine even asked to have canals wide enough to accommodate boats dug between the parterres so that she could sail directly from the river into her garden (this proved impracticable).[11] And once again all this variety was contained in rigorous geometrical patterns: the large square divisions, smaller square parterres within each, and elaborate geometrical arrangements of the flowers and vegetation within the parterres. Circles, triangles, oblongs, right angles, and diagonals were all amply represented. Even in the bosquets—parterres planted as "little woods"—the trees were carefully arranged in straight lines and contained within precise rectangles.

Founded by the great-granddaughter of Alberti's friend and patron, the Tuileries were an Albertian garden through and through. In their richness they represented the fantastic variety of the kingdom, and, in fact, of the entire world. In their regular patterns, they revealed its geometrical organizing principles, a hidden but all-pervading order that structures everything. In the Tuileries the world was revealed as a peaceful and harmonious place in which everyone and everything had their natural and inescapable place in the grand scheme of things. It was, in other words, everything that France in the age of Catherine de Medici was not.

Apart from its size and complexity, the Tuileries also offered one critical innovation when compared to preceding generations of royal gardens: there, for the first time, the gardens extended directly from the palace. At Amboise, the relatively modest garden was enclosed within the castle walls; at Blois it occupied an area near the main château; and Fontainebleau's numerous gardens more or less encircled the palace in an irregular pattern. But the palace and gardens of the Tuileries were inseparable: coextensive with the building's façade, the gardens began at its doorstep, the two forming a single unified complex. Anyone looking outward from the palace would see the garden, and little else. Conversely, looking up from the gardens, one's eye would inevitably be drawn to the palace that towered over the parterres, bosquets, and fountains.

For the prince in the Tuileries Palace the garden was his natural and peaceful domain; and for anyone walking the garden paths, the prince in his palace was the natural, inescapable overlord. By integrating the palace with the gardens, the Tuileries introduced a new element into Alberti's elegant geometrical harmonies: a strict hierarchy, with the monarch at its apex. To residents and visitors in the Tuileries, royal supremacy over an ordered and peaceful land was not an empty boast, but a truth manifested all around them.

Catherine was not slow to take advantage of her creation to teach her people and the world a lesson: in 1573 an artificial mountain was erected at the garden's center for a grand feast in honor of the Polish envoys who had come to Paris to offer the crown of Poland to Henry of Anjou, the future Henry III. As the guests arrived, sixteen of Catherine's ladies-in-waiting were seated on the mountain, each dressed as a nymph and representing one of the provinces of France. After reciting verses in praise of France and the future king of Poland, the nymphs descended into the garden and performed an intricate ballet. They concluded the evening by presenting gold medallions to the king, the queen mother, and the Polish guests. It was a stirring occasion that brought out the meaning of the splendidly diverse but carefully regimented garden. The precise geometries of the dance and the perfectly manicured parterres suggested a vision of national order and unity, where the provinces of France come together in harmony under the benevolent gaze of royalty.[12]

When the Tuileries celebrations took place in 1573, the French monarchy had sunk to one of the lowest points in its long history. The memory of the bloody St. Bartholomew's Day's Massacre was still fresh, and the victims had barely settled in their graves; the Huguenots had retreated to their provincial strongholds and vowed to defend themselves against what they considered a murderous monarchy; the Duke of Guise and his brother the cardinal were victorious at court but widely hated by the people, a hostility that easily transferred to the person of the king; the royal armies were making no headway against the Huguenots, and Charles IX's own brother, the Duke of Alençon, had joined forces with their leaders against the king. With the king a virtual prisoner of the Guise brothers, his authority did not extend even to his own court. To his subjects he was at best a well-meaning bumbler, at worst a wicked persecutor.

And yet it was precisely at this time, when the king's power and prestige were at their lowest ebb, that the queen mother decided to invest extravagant amounts of the monarchy's dwindling resources in creating the Tuileries and producing a grand public spectacle. It was a sign of the value Catherine placed on magnificent display as a means of preserving the beleaguered monarchy. But it was also a sign of how intertwined the faith in the monarchy had become with its geometrical gardens. The Tuileries presented an ideal kingdom that reality outside the garden walls could not match. They were, to be sure, a promise of a better future, but they were much more than that: they were a bold claim that monarchy, hierarchy, and order were the only natural and harmonious state of affairs, as universal and incontestable as geometry itself. Even though the brilliance of royalty might be temporarily dimmed, for anyone visiting the Tuileries gardens there was no escape from this universal truth. The squares and triangles of the garden in the shadow of the palace whispered it, the dancing nymphs in the spectacle proclaimed it: there can be no order, no harmony, no peace in the land without a king at its head.

The Return of the Kings

The fortunes of the kings of France reached their nadir in the 1580s, when an uprising led by the Catholic League matched the ongoing revolt of the Huguenots, and Henry III met with a violent end. Yet the king's assassination in 1589 turned out to be not only the monarchy's lowest point, but also a turning point. Lacking a son or even a brother to succeed him, Henry bequeathed his crown to a distant relative, his ninth cousin once removed, the same Henry of Navarre who had been his ally in his war on the League. And Henry IV, as he was now called, the first king of the house of Bourbon, proved to be the man who would begin to reverse the monarchy's decline, end the civil war, and finally bring (relative) peace to the land.

In some ways Henry IV merely continued the policies that were set in motion in the final years of his predecessor. Like Henry III before him, the new king took up the leadership of all the factions that were opposed to the extremists of the League, be they Huguenots seeking toleration by the state or Catholic "politiques" who had concluded that the only way to bring peace was to reach a practical compromise. At the head of this broad coalition Henry defeated the League's armies and once again laid siege to Paris. At long last Catherine's determination to keep the spark of the monarchy from being extinguished in the darkest days of the civil war was paying off, as Frenchmen rallied to their king. The ineptness of the king's enemies helped as well: as the war dragged on, the murderous excesses of the League, and its reliance on military assistance from the hated Spaniards, steadily eroded its appeal and increased that of the king. The final hurdle was religious: Henry was a Protestant, and even moderate Catholics who despised the brutality of the extremists and longed for the steadiness of royal rule were reluctant to accept a heretic king. And so in 1593 Henry abjured his faith and converted to Catholicism. The following year he was crowned in a lavish ceremony at Chartres Cathedral. Shortly thereafter, the governor of Paris opened the city gates and surrendered to the king.[13]

The war was not yet over, but its outcome was no longer in doubt. Over the next four years Henry systematically reduced the last strongholds of the League in the countryside, cementing his victory and his rule. Then, in 1598, he signed the Edict of Nantes, which defined the place of the Huguenots in France. While permanently confining them to the status of a small minority in the kingdom, the edict also allowed them to control certain territories, retain armed forces and strongholds, and establish official committees that would negotiate with the king in case disputes arose. It was a compromise that pleased no one, but the people of France were weary of war. With the edict signed by the king and registered into law by the Parlement of Paris, the civil war that had ravaged France since the death of Henry II came to an end.[14]

The end of the war did not exactly mean peace. For the rest of his years Henry still had to suppress repeated uprisings by provincial aristocrats who had grown used to a degree of local power and autonomy that the king would no longer tolerate. He had to deal with the armed presence of the Huguenots in his kingdom, which even in the best of times posed a threat to royal rule. And times were not always the best: in 1602 the king's friend and ally the Duke of Biron was executed for plotting against the king, and two years later the Duke of Bouillon, leader of the Huguenots, was implicated in a similar conspiracy. Finally, in 1610, after twenty-one years on the throne, Henry was assassinated by a Catholic fanatic who stabbed him in his coach, proving that while the League was dead, the passions that had animated it were still very much alive.[15]

Yet despite the continuing violence, and even the murder of a king, royal rule continued unchallenged. The sudden death of Henry II in 1559 had plunged the land into civil war, and the assassination of Henry III thirty years later had brought the monarchy to the verge of collapse. But when Henry IV was assassinated the crown passed seamlessly to his nine-year-old son, who became Louis XIII, and the queen mother, Marie de Medici, ruled as regent until the king came of age. The same thing happened when Louis in turn died in 1643, leaving his crown to the five-year-old Louis XIV, and his authority to Louis XIII's wife, Anne of Austria, who served as regent during the king's long minority. Kings and regents could come and go, occasional plots and regional uprisings persisted, even foreign invasions recurred periodically, and yet through it all the monarchy continued unchallenged. Kingship, which had teetered on the verge of irrelevance in the days of Catherine de Medici, had returned as the one institution France could not do without.

Henry IV's popularity stemmed in no small part from widespread disillusionment with his enemies, as it became clear that support for any of the rebellious factions meant support for interminable civil war. Yet the king also knew that if he was to prevent the country from reverting to a chronic state of war, this was not enough. France in 1600 was a land in which Catholics opposed Protestants, towns opposed the countryside, great aristocratic clans were at odds with one another, and the ancient nobility felt threatened by a rising class of officials and artisans. Forty years of conflict had resolved nothing. If the kingdom were simply to revert to the way it was in the reigns of Francis I or Henry II, it would take only a spark to reignite the country and set France once more on the path to war.[16]

And so, to save the monarchy and preserve the peace, Henry IV set about reestablishing the ancient institution on new foundations. The old monarchy acknowledged and cultivated the various orders and divisions in society. It granted special charters to different towns, special privileges to different trade guilds, special (if limited) rights to religious minorities such as Jews, and so on. Most significantly it acknowledged the rights of noblemen to rule over their domains without outside interference. All of these competing groups and interests could be kept in line because they acknowledged the king as the highest power in the land. In return, the rule of the kings over their subjects was mediated through institutions representing the different segments of society. Royal decrees could be passed down only through a complex hierarchical chain, and if the king had any hope of his will being done, he had to take into account the interests of the corporate bodies who would be putting them into practice.

Coming out of the civil war, however, Henry knew that a return to the ways of old would spell disaster. Acknowledging the power of the different corporate groups and the semi-sovereign status of the great nobles would mean handing back power to the groups most responsible for the kingdom's disintegration. And so, he determined, the new monarchy would be different: instead of relying on intermediaries to exercise royal rule, the kings of France would henceforth rule their country directly. Nobles great and small, city officials, and trade corporations would all be sidelined, as the power of the king would penetrate to every corner of the land and reach each and every subject directly.

It is hard to say how clearly Henry IV himself perceived these ideas. As Henry of Navarre, before his succession to the throne, he himself, after all, had been the leader of a rebellious aristocratic faction, and fellow warrior-aristocrats were his closest companions throughout his life. He never said "L'état, c'est moi," as his grandson, Louis XIV, reputedly did, and he never publicly expressed his desire to suppress the power of the nobility and corporations. But there is no question that he set in motion the policies of concentrating all power in the hands of the monarchy and centralizing all decision making in the royal court. They were policies that would define the French monarchy throughout the seventeenth century, and they had their origins with the first Bourbon king, Henry IV.

How did the Bourbons set about concentrating and centralizing power? For one thing, they simply settled down. Francis I, despite promising to settle down in Paris, had incessantly crisscrossed the kingdom to assert his authority with his nobles, and so did his successors. But Henry IV settled permanently with his court in the Louvre, which remained the primary royal palace until Louis XIV's move to Versailles in 1682. It was a bold move at a time when the monarchy was still weak, and challenges to the king's authority in different corners of the land often required his personal presence. But it was also of immense symbolic value: henceforth, instead of the king traveling to the domains of aristocrats to establish his power and legitimacy, they would travel to court to establish theirs. All power, it proclaimed, dwelt with the king in his capital, and anyone seeking to partake of that power must make his way to the royal court. From now on all eyes in France would be on the king's court, the sole source of patronage and largesse, as well as of swift, unforgiving punishment, as the king might see fit.

Symbolism, however, wasn't enough. In reality, the court in 1600 depended heavily on local authorities, be they nobles or city leaders, for tax collecting and for the enforcement of its decrees. This inevitably gave provincial magnates great leverage in their dealings with the court, and made it difficult for the king to defy them. To get around this, Henry began sending his own commissioners to the provinces, armed with broad authority and appointed for an unlimited duration. Under Louis XIII and Louis XIV the system was formalized, with every province receiving its own royal "intendant," responsible for tax collection, policing, and the administration of justice.

The intendants, unlike the traditional officials, owed all of their authority to the king and nothing to the local power brokers in their assigned territory. Each intendant, in turn, delegated his authority to subordinates, so that in the end every village and town in France was supervised by a representative of the royal bureaucracy. Since the old system was never officially dismantled, the result was the creation of a parallel administration alongside the traditional institutions that depended on the provincial elites, and in particular the nobility. When disputes inevitably arose, the intendants did not hesitate to call upon royal troops to enforce their authority. Over time, even in the face of determined opposition, the intendants sidelined the traditional institutions, overcoming the entrenched power of the aristocracy and corporations. Centralized royal power, an aspiration of kings dating back to the civil wars, was gradually becoming a reality.

Yet even as their economic and administrative power was systematically eroded, great nobles and independent-minded cities still held fast to their mighty strongholds. As long as nobles and townsmen could resist the monarch's decrees by hiding behind their fortifications, royal authority would be limited. To make things worse, the Edict of Nantes gave Protestant communities the right to keep their own strongholds and armed forces, and while this was a necessary concession, it also made the Huguenots the greatest of all threats to royal authority.

And so the Bourbon kings embarked on a campaign to demilitarize the countryside by dismantling both castles and city walls. After a slow beginning under Henry IV, ever cautious not to rekindle the embers of the civil war, the program gathered momentum under his successors. First Louis XIII and then Louis XIV systematically demolished thousands of castles and urban fortifications, forever altering the landscape of the French countryside. When the Huguenots rose up in 1621, Louis XIII seized the opportunity to bring them to heel as well, ultimately laying siege to their last stronghold of La Rochelle. When that Atlantic port surrendered to the royal army in 1628 it marked the end of Huguenot armed power.[17] Reign by reign, Henry IV and his successors were making good on their promise to nullify the power of all rivals to the monarchy, be they noble, urban, or religious.

By the middle of the seventeenth century the Bourbon kings of France had entirely transformed the desperate kingdom they had inherited in 1589. Within a few decades Henry IV and his successors managed to pacify the land, neutralize the independence of the great nobles, and crush the entrenched power of the Huguenots, and were well on their way to creating a centralized royal administration. It was a remarkable turnaround: the ravaged kingdom of the 1580s had reemerged as a European superpower and a model of effective royal rule.

Gardens for a New Kingdom

It didn't come easy. Henry IV was assassinated by a Catholic fanatic; Louis XIII fought an eight-year war against the Huguenots; Louis XIV spent his early years a virtual captive of the uprising known as the Fronde. And from the 1630s onward, when France intervened in the Thirty Years' War, foreign war became the norm in the kingdom's relations with its neighbors, extracting a heavy price in both blood and money. By the end of Louis XIII's reign France was fielding enormous armies a quarter of a million strong, and the crushing burden of taxes required to support them led to increasing resentment of the king's intendants and, on occasion, violent resistance to their impositions.[18]

Even faced with a restless countryside and depleted treasury, however, the kings of France never held back on what had become their signature obsession: lavish investment in geometrical gardens. Henry IV, known as the Gardener King, revamped the gardens at Fontainebleau, which had been neglected as the monarchy struggled to survive in the final decade of the civil wars.[19] In his most ambitious venture, he commissioned an enormous garden sloping down toward the Seine from the Château-Neuf at Saint-Germain-en-Laye west of Paris. Begun around 1595, the garden as shown in an engraving consisted of eight massive terraces arranged one above the other, from pools and canals in the water-themed terrace on the riverbank to elaborate parterres around the château at the top of the hill.[20] Yet for all its ambition, the garden at Saint-Germain-en-Laye was fundamentally flawed as a representation of royal power. The geometries of each terrace were, in all likelihood, elegant and refined, but they were insulated from each other and from the Château-Neuf. Together they pointed at nothing in particular, and if they were in some fashion under the sway of the royal palace at the top of the hill, it appeared to be a complex and heavily mediated relationship.

This left the Tuileries—newly repaired, revamped, and expanded by Henry—as the exemplary geometrical gardens of the early Bourbons. Though not as grand or innovative as the gardens of Saint-Germain-en-Laye, they were far more effective in conveying the monarchs' overriding message. The orderly geometrical gardens of the Tuileries, emanating from and pointing to the royal palace, were the ideal Bourbon state in miniature. As if to emphasize this point, Henry expanded the walls of Paris to encompass the palace and gardens, which had previously been just outside the city. The Tuileries were now an integral part of the capital of France, permanent seat of its monarchs, the focus of all eyes in the kingdom and the source of all power.

When Louis XIII (1601–1643) came to the throne at the age of nine he was already showing himself a true heir to the Gardener King. "I know how to make fine gardens," he wrote his father from Fontainebleau at age five, and his physician, Jean Hérouard, confirmed that this was no idle boast. The prince, he noted in 1607, "draws out a square garden in an *allée*, trenches and arranges it, and plants it with cabbage." A year later he recorded that Louis "goes to his little garden; amuses himself by digging; hands out tools to others saying: 'Work, or I will beat you.'"[21]

But it was not, in the end, the king who created the most lasting and impressive gardens of his reign. That honor was reserved for his mother, Marie de Medici, who served as regent during his minority, and to Louis's chief minister, the cardinal Richelieu (1585–1642).[22] Like Catherine before her, Marie was a Medici princess who grew up in Florence and became queen of France through a political marriage. Her uncle, the Grand Duke Ferdinand I of Tuscany, had supported Henry IV both financially and politically in the turbulent years after his accession, and facilitated the king's conversion to Catholicism and reconciliation with the pope. Ferdinand also interceded with the Holy Father to annul Henry's marriage to Marguerite of Valois, thereby making way for Marie, who married the king in 1600.

Queen of France though she was, Marie sorely missed the Pitti Palace in Florence, where she grew up, and the adjoining Boboli Gardens. And so in 1611, while serving as regent for her young son, she wrote to her aunt Christina, Grand Duchess of Tuscany, to ask for her help in re-creating her childhood home in the French capital. Rather than wait for her aunt's answer, Marie dispatched Louis Métézeau, one of her court architects, to Italy to study the Florentine palace. When he returned, measurements in hand, she immediately set him to work on her new Parisian residence, sparing no expense. It would be built on a property she purchased from a certain François de Luxembourg, and so it is known to this day as the Palais de Luxembourg.[23]

Then as now, the Luxembourg gardens were located south of the palace, with a large, elongated parterre emerging from the southern façade and stretching southward from there. The *parterre de broderie*, as this part of the garden was known, was shaped like the ground plan of a cathedral, with a large rectangular area topped by a semicircular "apse." It was effectively an extension of the palace and was, in effect, a garden in itself. Ranged along a clear north–south axis centered on the main entrance of the palace, it was divided into elegant geometrical shapes of circles, semicircles, squares, and crescents, which joined together to create the overall symmetrical shape of the parterre. Each of the segments was in turn gardened into elaborate geometrical patterns with such precision and finesse that they appeared "embroidered," thereby giving the parterre its name.

When considered in isolation, the Luxembourg parterre de broderie is not only a beautiful piece of gardening but also an effective exercise in royal propaganda. Though not as large as the Tuileries gardens, the parterre too is a geometrical garden that emerges directly from the château and is dominated by it. In some ways the subordination of the garden to the palace is even more pronounced here: whereas the Tuileries gardens were arranged into regularly spaced rectangles, which appear the same from any direction, the parterre at the Luxembourg gardens had a clear central axis that emerged from the château, an axis marked by a broad path and successive circular flower beds and pools. All eyes from the garden are turned toward the château, and the royal gaze is drawn ineluctably from the palace to the gardens. Like Catherine's famous garden, the parterre was a rich, ordered, and harmonious land, at peace under royal power and protection.

Though much transformed and, perhaps, diminished, the parterre de broderie is still the central feature of the Luxembourg gardens. Yet today's Luxembourg Park, which includes little beyond the parterre itself, is but a small part of Marie de Medici's garden of the 1620s. Instead of being an extension of the palace façade, the garden stretched another quarter mile to the east, and a full half mile beyond the parterre to the west. These spaces were, to be sure, arranged in regular rectangles and squares, and planted with geometrically patterned flowers and trees. But the palace, relegated to the eastern end of the estate, was largely irrelevant to most of the gardens, which consequently conveyed no message of overall hierarchy and harmony. The ideal of royal supremacy, powerfully presented by the parterre de broderie, was diluted to the point of irrelevance by the larger garden with its whimsical irregular overall shape. The Tuileries gardens, on the opposite bank of the Seine, were still the emblematic royal gardens of France.

A more sustained effort at representing hierarchy and rule through gardening was attempted by Louis XIII's chief minister Armand-Jean du Plessis, universally known as Cardinal Richelieu. Born to the high aristocracy in the province of Poitou, Richelieu was initially a favorite of the queen mother, Marie de Medici. His loyalties, however, proved flexible: when in 1617 the young king ousted his mother and her favorites from the court, Richelieu slipped seamlessly into the king's camp and soon became Louis's most loyal and capable servant. From 1624 to his death, as the king's chief minister, he was the chief architect of "Louis the Just's" policies of administrative centralization at home and European domination abroad. Other than the king himself, there was no man in all of France grander and more powerful than Cardinal Richelieu. And sometimes, it appears, the cardinal was tempted to forget even that sole exception.

That, at any rate, seemed to have been the case in 1631 when Richelieu began a vast construction project on his family estate southwest of the town of Tours. The cardinal had obtained permission for the project from the king, but one must wonder whether Louis was fully aware of the scale and ambition of his chief minister's designs. For the complex at Richelieu included not only a large château and a vast geometrical garden, but also an entirely new town, built from scratch to serve as an administrative center, complete with law courts, an academy, a printing press, and a religious mission run by the Sisters of Charity of Saint Vincent de Paul. What territory exactly was supposed to be administered was never fully spelled out, but the scale was certainly worthy of a kingdom. Indeed, the only other missions belonging to the Sisters were in the capital of Paris and the royal palace at Saint-Germain-en-Laye.[24]

The cardinal's estate was organized along an east–west axis with the château itself at the center. The western entrance was located at the edge of a massive circle 100 meters in diameter, where three broad avenues, bordered by elm trees, converged. In a pattern known as "trivium" (which would be repeated some decades later at Versailles), each avenue was ruler straight, and together they formed a giant arrow pointing toward the château. A visitor entering the estate would then pass through two successive courtyards surrounded by rows of administrative buildings before finally arriving at the main house. Passing through the house to the eastern façade, he would finally arrive at the gardens, composed of square and circular parterres with pools and fountains arranged symmetrically around a central canal. Beyond were the open woods, with a broad avenue extending the central axis of the château into the distance. The newly built town of Richelieu was located to the north of the château, shaped as a precise rectangle divided by regular orthogonal streets.[25]

The château and town of Richelieu was a geometrical construction, and it was designed to exude not only splendor, but also raw power. The straight grand avenues converging on the estate would show the visitor that all roads lead to this seat of power. The successive rectangular courtyards would impress upon the traveler the layers of authority one had to traverse before approaching the great lord, and the carefully patterned geometrical parterres told of the peaceful, harmonious land beneath the palace's sway. The scale of the courtyard and gardens and the open horizons to both east and west carried the suggestion that the ordered lands and the domain of the lord are unbounded, extending to infinity. Most of all, the rectangular town with its orthogonal streets is a sign of order and rationality harnessed to the service of the palace and the land. Created from nothing by the lord's diktat, the town is a testament to his near godlike power and resources.

Grand and magnificent as it was, it is hard not to see Richelieu's estate as a challenge to the ascendancy of the king. Nor was the country estate the cardinal's only venture into large-scale landscaping. In the very heart of Paris, just north of the Louvre, Richelieu built the "Palais Cardinal," complete with an elegant geometrical garden. The palace was much smaller than the Louvre, and its park less than a quarter the size of the Tuileries. And yet their presence so close to the seat of royal power was an act of hubris, if not impudence, for it suggested that there was not one center of power in Paris, but two. Louis XIII, for all that, seemed unconcerned with his minister's pretensions, and ultimately his confidence proved justified. The Palais Cardinal remained Richelieu's Parisian headquarters until his death in 1642, when it was handed over to the king and became the Palais Royal, as it is known to this day. With this larger-than-life minister out of the way, the two centers of power in Paris were easily folded into one.

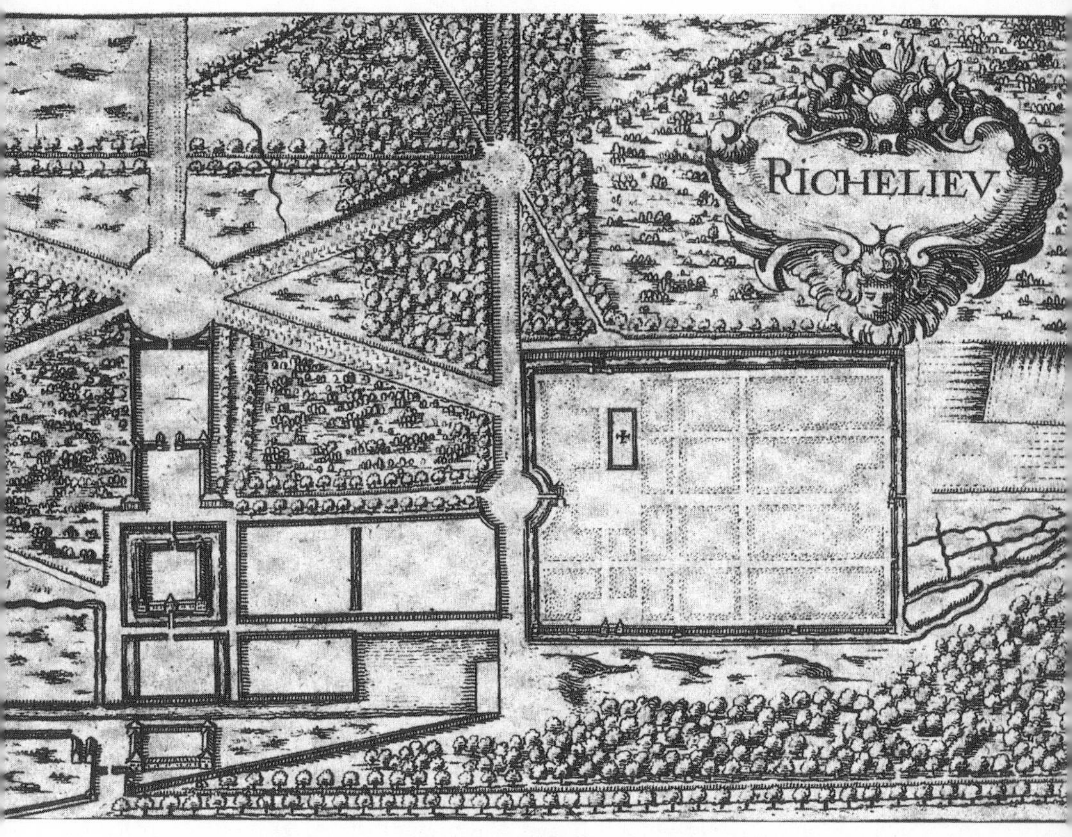

FIGURE 15: **The garden and town of Richelieu in the seventeenth century**

The great estate near Tours, however, was a different matter, both because of its vast scale and ambition and because it became the family seat of the cardinal's successors the dukes of Richelieu. If Louis XIII nevertheless took no action against his minister, perhaps it was because the château and town were so far removed from Paris and the court that the cardinal's grand pretensions to rule over an empty countryside seemed a tad farcical. Perhaps it was because Louis never saw the estate, for Richelieu—unlike the imprudent Fouquet some decades later—did not invite his king to celebrate the splendor of his castle with him. Perhaps Louis trusted Richelieu's unconditional loyalty, or perhaps he was so dependent on him in managing the affairs of state that he could not risk the rupture that would ensue from disciplining the cardinal. Or perhaps Louis was mollified by the giant statue of himself on horseback that the cardinal had prudently erected in the central courtyard at Richelieu. We will never know the reason for the king's tolerance. Yet if Richelieu had been serving Louis XIV rather than his father, no equestrian statue would have saved him from the king's wrath.

The Noble Gardener

The creation and maintenance of the monarchy's vast and sophisticated gardens inevitably required a small army of professional gardeners, from simple workers charged with planting, trimming, and watering to designers and architects, learned men with powerful positions at court. Mercogliano was, so far as we know, the only man charged with the design and creation of the garden at Amboise in the 1490s. A century later, however, entire dynasties had been established whose members bore such titles as *jardinier ordinaire du roi* and even *intendant des jardins de roi*, and passed on the titles—as well as the profession—from generation to generation.[26]

An indication of just how central the gardens had become to the life of the monarchy is the fact that by the turn of the seventeenth century gardening had become an attractive career path not just for artisans, but also for members of the nobility who were seeking a secure and prestigious place at court. Such a man was Jacques de Boyceau de la Brauderie (ca. 1560–1633), born to an ancient family from the Saintonge region, not far from La Rochelle. As members of the minor nobility, the Boyceaus attached themselves to the Huguenot magnates of their region, the dukes of Biron, and prospered along with them. Jacques's father was a distinguished member of the household of Duke Armand de Biron, who was made Marshal of France by Henry III, and Jacques himself was a close friend of his son, Duke Charles de Biron. Together they fought in the armies of Henry of Navarre, and they continued in his service when he became Henry IV of France. As a reward for his services, Henry named the young Biron admiral de France, and subsequently marshal and peer of the realm. Jacques, the duke's brother-in-arms, basked in his friend's reflected glory, his place in court seemingly secure.

It didn't last. In 1602 Duke Charles was arrested for plotting against the king and executed in short order. It was a stunning reversal of fortune for the duke, and a clear sign that times had changed. Great nobles changing sides and plotting against one another and against their monarch had been commonplace during the decades of the civil wars. The only recourse for a betrayed monarch had been to try to win back their support through favors, or, failing that, assassination. But Henry would have none of it: anyone plotting against the monarch, he signaled, even a friend and favorite of the king, would be treated as a traitor and a criminal and dealt with accordingly. In the new monarchy the great nobles would either be loyal or they would be crushed.

With his friend and patron denounced and executed, Boyceau's position became precarious in the extreme. Suddenly any association with the duke, or even the mention of his name, was enough to put your life in danger. Yet somehow Boyceau managed to survive at court, and even to thrive. As a member of the lower nobility he was not viewed as a threat like his grand friend, and besides, the king was fond of his devoted follower who had fought in his battles for so many years. To signal his approval, and protect Boyceau's position in court, Henry appointed him to the position of ordinary gentleman of the king's chamber (*gentilhomme ordinaire de la chambre du roi*), a title he retained until his death.

This does not mean that Boyceau did not pay a price for the fall of his patron. Up to this point Boyceau had distinguished himself as a warrior, the proper vocation for a young nobleman. Yet any ambitions he may have harbored to pursue a military career now came to an abrupt end. The king, for all his goodwill, could not allow a close associate of a convicted traitor to command royal troops. Even if he was not concerned about Boyceau's personal loyalty, he was certainly concerned about the example such an appointment would set. And so the young courtier was forced to choose a different path, one that would make him valuable to the king while at the same time demonstrating his unswerving loyalty. He chose gardening.[27]

To Boyceau's peers it was a stunning choice. Gardening was a manual art, entirely unfit for a gentleman. Even master gardeners such as Mercogliano or members of the Mollet and Le Nôtre clans of the Tuileries, men who designed and supervised rather than dug and planted, were tainted by the association of gardening with manual labor. The proper relationship between a lord and a gardener, as everyone knew, was that between a master and his servant. Yet here was an aristocrat by birth, a member of an ancient noble family, a warrior and a distinguished courtier, who chose to specialize in an art that was clearly beneath his station. What was he thinking?

The answer is that Boyceau understood what many of his peers at court did not. Royal gardens, he realized, were not just frivolous spaces where the king and his courtiers could luxuriate, feast, and entertain. They had become central to the monarchy's public presentation and ideology, even its identity. By associating himself with a field that was key to the monarchy's power and legitimacy, Boyceau was demonstrating his unwavering devotion to the king and further distancing himself from the legacy of the traitor Biron. Even more important, Boyceau realized that the great gardens of Fontainebleau, Saint-Germain-en-Laye, and, most of all, the Tuileries were vital tools of policy, playing a critical role in Henry IV's reinvention of the monarchy as a streamlined centralized administration.

With a traditional military path barred to him, there was no better way for Boyceau to make himself valuable to the king than to recast himself as the leading gardener at court. His birth and court connections made him the superior of all other gardeners in the king's service, and he inevitably became the king's top gardening official, with the official title of *intendant des jardins du roi*. In this capacity he became a powerful presence at court, the indispensable link between the Sovereign's wishes and their execution in his gardens. The transition from warrior to gardener may have seemed to Boyceau's contemporaries to be the disgrace of a once-proud nobleman. In fact, it turned out to be a brilliant career move that extracted Boyceau from the cloud he was under and secured his position in the entourage of two kings.

As superintendent of the king's gardens, Boyceau undoubtedly played a role in the design and creation of the Luxembourg gardens. Although the paperwork of the garden's construction was lost in a fire in 1690, Boyceau's *Traité du Jardinage*, published posthumously in 1638, includes detailed designs of Luxembourg's famous parterre de broderie. Additional designs in the text are attributed to other parterres at Luxembourg, as well as to some at Saint-Germain-en-Laye and the first gardens of Versailles.[28]

The *Traité du Jardinage*, however, is much more than a technical manual about gardens and their creation. It is a public statement on the meaning and purpose of gardens, at a time when they were acquiring an outsize role in the presentation and functioning of the French monarchy. The book's frontispiece suggests as much when Boyceau is identified not as a mere gardener, but as "Esquire, Lord of Brauderie, Ordinary Gentleman of the King's Chamber, and Intendant of His Gardens." The *Traité*, in other words, authored by the highest-born and most illustrious gardener in France, is an all but official statement on gardening, as viewed from the heart of the new monarchy's court.

At its core, according to Boyceau, gardening is nothing but the imitation of God's creation: "Oh Lord," he calls out in a poem in the opening chapter, "whose word created this world from nothingness, who brought about the herbs and the flowers, the trees and the plants,"

Illuminate our sense, incapable of seeing
The marvelous roots of your divine power
Teach us the secrets of your daughter, Nature
So that we will follow her path in our agriculture . . .[29]

A garden is nothing but man's puny attempt to follow in the paths of nature, and thereby re-create the fecund richness of God's creation. It follows that the closer a garden gets to the prodigious variety of nature, the better. As Boyceau put it, since "Nature gives us such variety, we judge that the most varied gardens are found to be the most beautiful." Such variety in a garden, he continues, can manifest itself in many ways:

in differing locations, in the general outline of the garden, in the design of the parterres, and in the different plants and trees, with their range of shapes and colors.[30]

Yet richness and variety of the many elements in a garden must not be confused with chaos and disorder: for "all those things, as beautiful as can be chosen, are defective and less agreeable if they are not ordered and placed symmetrically and in good correspondence." Hence a proper garden must be carefully arranged according to strict guidelines of order and symmetry. For Boyceau, a garden in which trees, shrubbery, and flowers are allowed to intermix freely and flow into one another without careful design is not only lacking in beauty; it also fails in the essential purpose of a garden. It is hardly a garden at all.

To us, Boyceau's insistence that a garden must be carefully designed and ordered to be beautiful is at odds with his assertion that it is, at its core, an imitation of nature. We, after all, are used to thinking of nature as wild and unbridled, free of such human-imposed constraints as straight lines or precise symmetry. Nature, we believe, is to be found in places untouched by humans, whether in the dramatic desert landscape of the Grand Canyon, the lush greenery of a Brazilian jungle, or the open sea. But Boyceau would dismiss our view as superficial, a classic case of missing the forest for the trees. Nature, for him, is ordered, and gardens must observe symmetry and good correspondence precisely "because Nature observes them in her most perfect creations." "We can do no better," he wrote, "than to follow this great mistress in this, as in other particularities that we have touched upon." Order and symmetry in a garden are not imposed on it from the outside but are already there, hidden in the deep architecture of nature. The whole purpose of a garden is to make this hidden order visible.[31]

What kind of order is it that nature conceals and gardens reveal? It is, of course, geometrical order: "The square forms are the most common in gardens, whether perfect squares or rectangles," Boyceau begins, which was indeed the case in French Royal gardens from Amboise to the Tuileries. One of the advantages of such an arrangement is that it creates long, straight avenues, which "makes the things smaller and tending to one point." Such orthogonal shapes, in other words, re-create in the garden the geometry of linear perspective. The geometrical space of a painting, which Alberti described in *Della Pittura*, is now laid out in the three-dimensional garden, and with the same effect: all parallel lines converge toward a single vanishing point at infinity.

Yet squares, rectangles, and parallels are not enough to make a pleasing garden. "I do not advise," Boyceau opined, "that, making everything these straight lines, we neglect to intersperse rounds and curves; and among the squares, oblique lines." The reason is that a garden must contain a range of different patterns, "so as to find the variety that nature demands."[32] "I am greatly weary of seeing all the gardens partitioned only in straight lines, one arranged in four squares, another in nine, another in sixteen, and never seeing anything else," he complained, in an unflattering assessment of French royal gardens from the days of Mercogliano onward.

Far from recommending the abandonment of geometrical forms in favor of a free-form landscape, Boyceau's complaint was that the square arrangement was not geometric enough. What was required was a richer geometrical world that would include more-complex formations. In addition to the unimaginative squares, he continues, "the other perfect forms should also find their place and graces in the garden."[33]

Squares and rectangles; straight, oblique, and curved lines; triangles, hexagons, octagons, and pentagons—these for Boyceau are the building blocks of the perfect royal garden. It is a geometrical world through and through, in which all patterns, shapes, and angles are precisely laid out, leaving no room for whimsical or undisciplined departures. It is a world in which every tree, rock, pond, and blade of grass has its precise unalterable and nonnegotiable place in the grand scheme of things. It is the wondrous kingdom, peaceful, rich, and—most of all—orderly. It was the kingdom that the Bourbon monarchs were struggling fiercely to create, a land in which everyone had their God-given place, and all lived together in prosperous harmony under the benevolent gaze of the king.

To anyone walking the paths of Boyceau's geometrical gardens, the lesson was clear: the rule of the kings of France, their claims to absolute power, their centralized administration, and their intolerance of dissent were not a brutal imposition by a powerful prince upon a recalcitrant land and unwilling populace. These were, rather, part of the natural order of creation and, like geometry itself, could not be challenged or altered in any way. Through their gardens, the French monarchs became a central part of that natural order, rulers to whom any opposition was not only unwise, but also unthinkable.

For nearly two centuries, both Valois and Bourbon kings had labored mightily against heavy odds to realize their dream of a geometrical kingdom with themselves at its apex. None would come closer than the child who ascended the throne in 1643 following the death of his father, Louis XIII. To his admirers he would be known as the Sun King, the most glorious of French monarchs and the embodiment of absolute rule. To his detractors he remains a bloody tyrant with delusions of grandeur and dreams of universal empire. To history, he is Louis XIV.

5.

THE KING'S GARDEN

The Bedchamber

If you were traveling on horseback from Paris on an errand to King Louis XIV around the year 1700, you would ride into Versailles from the east, along the broad, tree-lined, and arrow-straight Avenue de Paris. With the end of this journey finally in sight, you might hasten your horse through the center of town and toward the gates of the royal palace beckoning at the end of the road. But before you got there, all would suddenly change: the lines of elegant houses would end abruptly and the avenue would open up into a large triangular plaza, broad at the base and narrowing toward the palace gates. Here you would notice two other ruler-straight boulevards sweeping like great rivers into the plaza: the Avenue de Saint-Cloud to the north, the Avenue de Sceaux to the south.

Together the three converging roads formed a giant arrow—the plaza and two outside avenues constituting its head, the Avenue de Paris its central shaft. It was pointed directly at the center of the palace.

The plaza where the great avenues converged was known as the Place d'Armes, for it was there that the king's household troops drilled and paraded. But on this morning the Swiss and French Guards and Royal Musketeers were in their barracks and the space was given up for the comings and goings of townspeople, courtiers, and officials. Hurrying across, you would present your credentials to the sentries at the gate and be admitted into the *avant-cour*, the outermost courtyard of the palace. Here the open space of the Place d'Armes becomes an enclosure, walled in by official buildings to the north and south and the shadow of the palace beyond. As you move toward your appointment with the king you proceed through two more increasingly confined courtyards, known today as the Cour Royale and the Cour de Marbre, surrounded by the towering stone walls of the palace. You are at the tip of the great arrow, and the power of the king, the court, and the state is closing in around you.

If your mission was sufficiently important, you would then be admitted and ushered into the king's bedchamber, located at the very center of the palace. In the middle of the room is the royal bed, aligned precisely with the central axis of the palace, which itself is a direct continuation of the Avenue de Paris. It is to the king in his bed that the arrow is pointing, for it is there that all lines meet, all roads converge, and all power resides. You are now at the center of the center of the kingdom and its living heartbeat, the place to which all the king's subjects look for guidance and from where his benevolent rule radiates to every corner of the land. Overwhelmed by the royal presence and the weight of symbolism in the room, you stumble through your presentation to the king, confusing the proper forms of address and etiquette that you had worked so hard to master. But on this day at least Louis seems more amused than angry at your uncourtly performance. He acknowledges your message and orders you to await his answer in the palace gardens. You bow low and exit the room without ever turning your back on the king.[1]

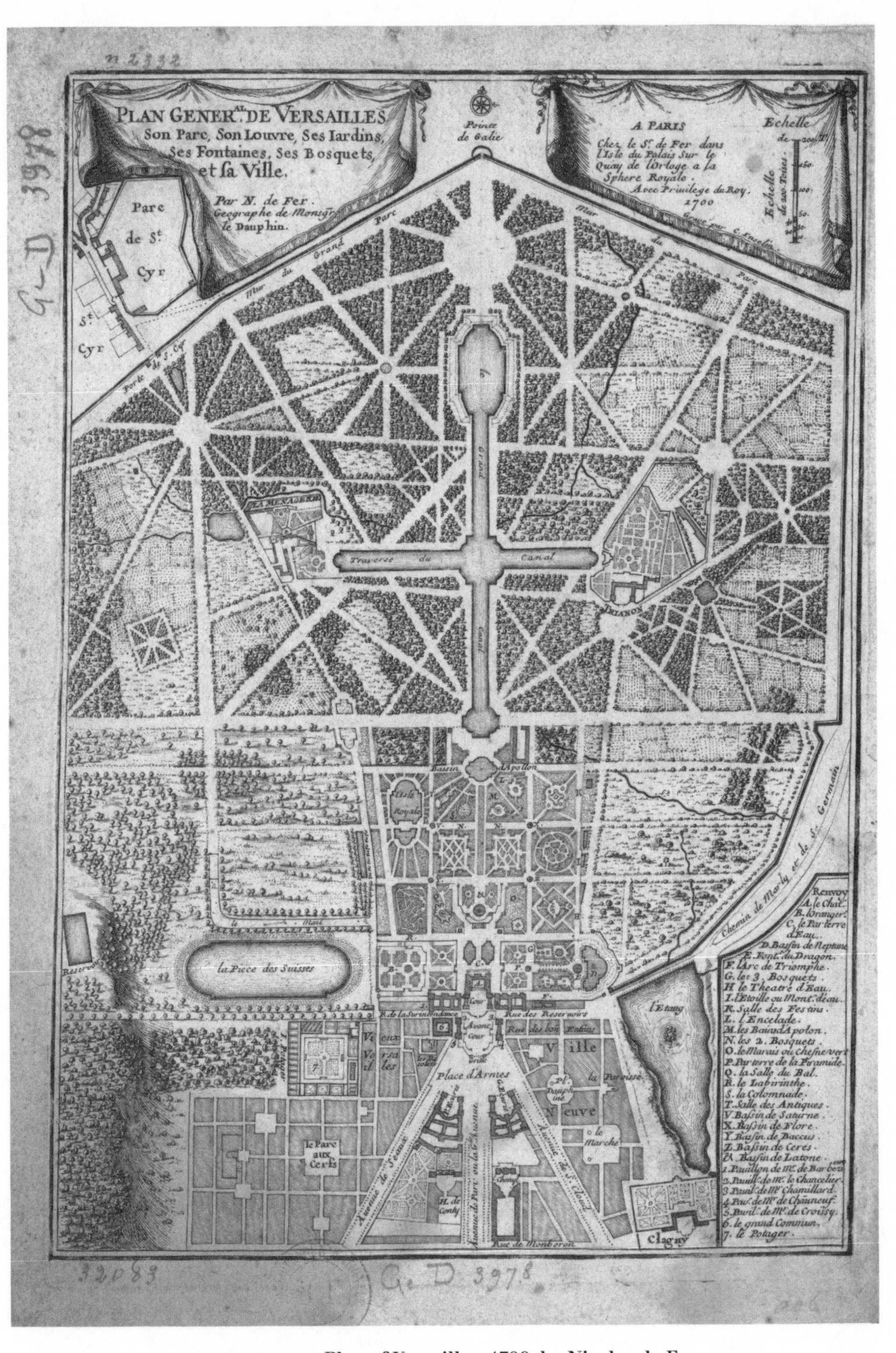

FIGURE 16: **Plan of Versailles, 1700, by Nicolas de Fer**

You are now led behind the bedchamber into the ground floor of the palace, where the light coming from three glazed doors reveals a scene of startling beauty. The confined space of the bedchamber and palace opens up at once to a world of brilliant colors and flowing waters, sloping down from the palace as far as the eye can see. If the château is an enclosed world of severe stone, the gardens are a magical landscape of flowers, shrubs, trees, and fountains of every conceivable kind reaching out to the horizon. Dominating the scene is the Grand Canal, whose main arm is precisely aligned with the central axis leading from the Avenue de Paris through the royal bedchamber, and the broad Allée Royale that descends from the palace, dividing the gardens into two symmetrical parts. Beyond, the valley is crisscrossed with a web of avenues, some broad, some narrow, but all of them ruler straight and intersecting at precisely measured angles. All ultimately lead to the great palace at the top of the hill.

Walking down the Allée Royale, you pass between flowery parterres and explore the wooded bosquets of the "Petit Parc," before descending to the banks of the Grand Canal. For all their richness and variety, you discover, the gardens are immaculately ordered, every part precisely arranged with respect to all the others. Nothing can be altered without destroying the overall design. Dominating all the disparate parts, visible from every corner of the gardens, the apex where all lines meet, is the palace, with the royal bedchamber at its heart. You can walk for miles along the open garden paths and never for a moment escape the royal gaze. The realization overwhelms you: the king is not just a powerful prince, but the centerpiece of an irrevocable geometrical order that encompasses all of nature.[2]

The simple geometry of the three great boulevards of Versailles told of how the eyes of all Frenchmen converged on the king in his chamber, and how power emanated from him to every corner of the kingdom. The gardens extended the vision beyond the world of men, even beyond France, and made royal supremacy part of the order of creation. In 1623 Galileo boldly claimed that the world is written in the language of geometry. In Versailles this was literally true. For there could be no better illustration of the principles of Euclid's *Elements* than the deep geometrical order that pervaded the royal gardens.[3]

At the apex of this universe was, inevitably, the king in his palace, whose rule was as inescapable and unchallengeable as geometry itself. To an outsider, Louis XIV's claims to absolute royal supremacy might sound like the ravings of an overambitious and power-hungry prince for whom no amount of power is ever enough. Many in Europe believed just that and never tired of denouncing and even mocking the monarch who unabashedly styled himself the Sun King. But to a courtier or a visitor to Versailles, one who had spent time in the palace gardens and walked their paths, those same claims were something else: a simple truth, self-evident and undeniable, rooted in the order of the universe.

The gardens of Versailles, with their embroidered parterres, straight avenues, and strict geometrical design, did not spring forth fully formed in the 1660s. They were, as we have seen, the culmination of a centuries-long tradition of French royal gardens dating back to Charles VIII's sojourn at Poggio Reale. Put differently, without Mercogliano, the itinerant Italian gardener brought to France by Charles, there could have been no Le Nôtre. And yet, as any visitor—whether in the seventeenth century or today—knows, Versailles is also qualitatively different from anything that came before it. The gardens of the Tuileries, Fontainebleau, or the Luxembourg palace are all exquisitely beautiful and even inspiring. But they simply cannot compare with the sheer awe, the overwhelming experience of having entered a different world, that greets a visitor to the gardens of Versailles.

First there is the matter of scale: the gardens at Versailles in the seventeenth century, including the hunting grounds of the "Grand Parc," were several orders of magnitude larger than anything that preceded them, extending to an astonishing 37,000 acres. Even the intensely gardened area seen from the palace, which is roughly equivalent to today's gardens, was huge, covering 1,890 acres. The entire area of the Tuileries, by comparison, was about 60 acres. When seen from the palace, stretching to infinity, the gardens of Versailles do indeed seem to encompass the world, not only metaphorically, but in reality.

Then there is the display of royal wealth and power, so ostentatiously presented in the gardens. For no one but the most powerful king in the world could command the workforce required for such a display. As any visitor to the gardens would suspect, building and maintaining this kingdom within a kingdom required a veritable standing army of workers that numbered in the tens of thousands. According to a well-placed courtier writing in 1685, there were no fewer than 36,000 artisans, painters, sculptors, and landscapers working in and around the gardens that year. No other garden, royal or otherwise, in France or elsewhere, had a workforce remotely approaching this scale.[4]

Then there is the matter of the gardens' design, and the careful arrangement of the various parterres and bosquets into a unified whole. Other royal gardens, and in particular the Tuileries, were designed as a succession of square segments laid out in careful rows, side by side, like open-air rooms extending from the palace. At Versailles, in contrast, the geometry is triangular rather than square, with all lines converging on the central, east–west axis that leads to the palace. Finally, the idealized land of the gardens is mirrored in the carefully designed town, whose grand avenues converge on the palace from the east, just as the broad avenues of the garden converge on it from the west. The message is clear: the same harmonious order that prevails in the wondrous gardens also orders the life of the very real, and often mundane, kingdom of France.

The overall effect on a visitor to Versailles goes beyond the geometrical order and harmony that permeated all royal gardens. It is also a sense of overwhelming power and rigid hierarchy that encompasses everything. Every single object in Versailles is part of a grand and inflexible chain of being: from a simple stone or blade of grass, through flowers and trees, fountains and canals, and even animals and people, all the way to the king in his palace. In their vastness, richness, and immaculate design, the gardens are not just an ideal kingdom. They are an entire ideal world, in which order and hierarchy permeate all, leading inexorably to peace, harmony, and transcendent beauty. In Versailles, this message was not elaborated in legal and philosophical treatises, or preached from on high by learned clerics and professors. It was, rather, a self-evident and encompassing reality, present everywhere one looked.

To the people who walked the garden paths of Versailles in the age of Louis XIV, be they resident courtiers or visitors from near or far, this notion of order was not only believable, but welcome. Memories of chaos, strife, and civil war were still fresh in their minds, and the establishment of absolute royal power carried not the threat of tyranny, but the promise of peace and prosperity. At Versailles they saw not the artificial designs of a well-funded propaganda machine, but the expression of a universal truth. Here the deep inherent order of the world was finally made visible. And it was, needless to say, geometrical.

The gardens of Versailles may be an expression of undying universal principles, as seemed clear to the courtiers who walked their paths in the seventeenth century, or they may be the outsize product of royal propaganda, as those who feared the king's power claimed even then (and many today would agree). Most likely they are both, a paradox combined in the vision of a single outstanding man. He was a man who did not hesitate to use elaborate artifice to achieve his ends, while also being a true and fervent believer in hierarchical order and royal supremacy. He was, like the gardens themselves, both the heir to a long and powerful tradition and an ardent innovator who departed from all that came before him. He was, of course, King Louis XIV.

The King at Bay

The birth of the child who was to be Louis XIV seemed nothing less than a miracle. His parents, King Louis XIII and his queen, Anne of Austria (1601–1666), had by that time endured twenty-three years of unhappy, childless marriage and had all but given up hope of producing a legitimate heir. But on September 5, 1638, at the age of thirty-six and to the astonishment of all, the queen gave birth to a healthy boy at the royal palace of Saint-Germain-en-Laye. He was named, accordingly, Louis Dieudonné—Louis the gift of God.

We know little of young Louis's earliest years, since, unlike his father, he did not have a companion such as Jean Hérouard to record his every move. We do know that he did not see much of his father the king, who died before he reached the age of five. But he was attached to his mother, forging a bond that lasted until her death in 1666 and was one of the closest of his life. He was a serious boy who from a young age was aware of his exalted place in the world, even as he was unceremoniously shuffled from palace to palace and often kept company with the children of servants. In April of 1643 he was brought to his father's deathbed in the palace of Saint Germain. "Who is it?" asked the dying king; "Louis XIV," replied his son.[5]

The young dauphin had anticipated his elevation by but a few weeks: on May 14, 1643, the old king breathed his last, and Louis Dieudonné did indeed become Louis XIV of France. King though he was, Louis was not yet five years old, and had many years of minority before him, rife with dangers to his life, his lineage, and the stability of his kingdom. His father did not help matters when in his will, drafted in his final days, he appointed a regency council to rule after his death, which could not be dismissed until the young king reached majority. The plan was a direct swipe at his wife, who in line with tradition was expected to be handed the reins of power, but as it turned out, the dying king need not have bothered. Anne took the child king before the Parlement of Paris and had the leading men of the kingdom, including those named to the regency council, proclaim that the law required that she be named sole regent. The Parlement agreed, and Anne of Austria, like Catherine de Medici and Marie de Medici before her, became regent during her son's minority.[6]

Though officially the sole regent, Anne did not rule alone. As her chief minister, and effective co-ruler, she chose Cardinal Jules Mazarin, an Italian from an old Sicilian family who had been Richelieu's right-hand man and anointed successor. In some ways Mazarin proved a worthy heir to his mentor. Possessed of sharp intelligence, ruthlessness, and unmatched political instincts, he put his steely ambition into the service of the regency and the state with impressive results. Together with the queen mother he continued the centralizing policies of their predecessors, enhanced French power in Europe, and preserved Bourbon rule through the turbulent years of Louis XIV's minority. Yet while

Richelieu had been feared and was hated by those who suffered at his hand, Mazarin was detested and despised to an extent that could not but hurt the standing of the monarchy. That a foreigner had become the co-ruler of France offended French national pride, and that he had done so through his personal influence on the queen mother encouraged every kind of tawdry speculation. As a result, Mazarin's presence at the pinnacle of power in France was an ongoing scandal that gnawed at the foundations of the regency.

Tawdry they may have been, but the persistent rumors about the queen mother and the cardinal were also, in all likelihood, well founded. Even the widespread speculation that the two were secretly married may have been true, since Mazarin, though a cardinal, was not a priest and had never taken a vow of chastity. The evidence is inconclusive, and in any case, to Louis XIV the legal technicalities hardly mattered. Married or not, the cardinal acted toward him with the cold authority of an overbearing stepfather.[7]

As the queen's partner, Mazarin was in charge of the royal household, including Louis's upbringing and education. Although the cardinal was initially uninterested in the child king, his mentorship was not without its advantages. As the king matured, the cardinal gradually took him into his confidence, showing by example rather than formal instruction how to conduct the affairs of the state. In all of Europe there was no more effective statesman or greater master of the art of politics, and the young king watched and learned. "I did not fail," Louis recalled years later, "to test myself in secret, reasoning alone and to myself about all events that arose, full of hope and joy whenever I would discover that my first thoughts were the same with which able and experienced people finally concluded."[8]

But overall, and despite this valuable apprenticeship, Louis recalled his years under Mazarin's authority with unalloyed bitterness. First there was the matter of his formal education, which suffered grievously under the cardinal's neglectful regime. Unlike other young nobles of his age, Louis never received proper humanist instruction, and to the end of his days he had to cover up his deficiencies in the ancient languages and literature that constituted the shared inheritance of the educated classes. Then there were the humiliating financial constraints imposed on the king, since although Mazarin managed to amass an enormous fortune for himself and his family, he kept his sovereign in a state of dire poverty. Even as the king reached majority and the regency was officially at an end, he still had to ask for the chief minister's approval for even the most trivial household expenses. In public he was the illustrious king of France, with all the pomp and magnificence due to his royal person. In private he had not a sheet without holes and was required to dismiss his servants to save the cost of their food. It is easy to imagine the effect this contrast had on the proud young king. There would come a time when no man would dare subject him to such indignities.

Yet the worst humiliations suffered by Louis in the early years of his reign came not at the hands of Mazarin, but at those of his enemies. The centralizing policies of the Bourbons were deeply resented by local magnates who saw their power and influence steadily eroded. Especially loathsome to them was the monarchy's attempt to rule the towns and provinces through intendants, who owed all their power to the king, and nothing to the established aristocracy and bourgeoisie. With the monarchy weakened by a prolonged minority, and the mighty Richelieu succeeded by the detested—if wily—Mazarin, the traditional elites saw an opportunity to strike back and regain their power and standing. In 1649 they launched a revolt against the regency that lasted four years and brought back memories of the dark days of chaos and civil war. Named after the slings used by the crowds to smash the windows of their royalist enemies, it became known as the Fronde.

The "Frondeurs" were far from a united front. The early years of the revolt were led by the urban Parlements, the powerful courts of law that were tasked with registering the king's decrees into law. Chafing against Mazarin's determination to turn them into rubber stamps of royal authority, they led a popular uprising against the oppressive rule of the royal bureaucracy. In an effort to bring them to heel, Mazarin turned to the great nobles, and in particular to the king's cousin Louis, the Prince de Condé, who was also France's greatest military commander. But when Condé, as a result, emerged as a powerful rival to the chief minister, Mazarin turned on him as well, eventually driving him to take up arms against the king in alliance with France's bitter enemy, the king of Spain.[9]

Through it all the king was effectively Mazarin's hostage, ushered from Paris to the provinces and back again in accordance with the cardinal's fluctuating fortunes and convoluted designs. The lowest point for Louis came in 1651, when he and his mother were kept prisoner in the Louvre by an armed militia of the Paris citizenry. Acting on information that their royal captive might have escaped, the citizens' guard demanded to inspect his quarters, and Anne, powerless to resist, agreed. This led to the outrage of commoners entering the royal bedroom and observing the young monarch as he pretended to sleep. For someone like Louis, raised to believe that kings were gods on Earth, the humiliation was unbearable. The poverty in which Mazarin kept him was an offense to his person, and one that he would not forget. But the presumption of the Frondeurs was also an offense to his station, and therefore one that he could not forgive. Louis would reign for another sixty-four years, every day of which was devoted to ensuring that commoners would never be allowed to take such liberties with their monarch again.[10]

By the end of 1653 the revolt had been put down, and Mazarin was once again firmly in control, pursuing the same repressive policies with renewed vigor. He crushed any hint of provincial autonomy and privileges, overruled local elections for town and village officials, and reintroduced the intendants, who had been withdrawn during the rebellion. To revive the kingdom's depleted finances he called on two loyal followers who could be trusted to ruthlessly extract the sums he required. One was a crusty old diplomat named Abel Servien; the other was the young, brilliant, and courtly Nicolas Fouquet, who soon began work on his dream château of Vaux-le-Vicomte.[11]

And just as the cardinal's authority was restored in the land, so it was in his own household. In 1654 Louis was crowned in a grand ceremony at Reims Cathedral, yet he still deferred in everything to Mazarin, outwardly expressing nothing but love and admiration toward him. He seemed to have no more control over his life, not to mention his kingdom, than he had had as a child. He fell in love with the cardinal's niece, the black-eyed Marie Mancini, in what may have been the one disinterested passion of his life, but it was to no avail. Whereas the chief minister may have been amenable to the match, it was ironically his purported wife, the queen mother, who threatened to raise an army to prevent it. Louis, predictably, gave in, and dutifully married the Infanta of Spain in June of 1660.

Courtly, respectful, and obedient to his elders, Louis XIV gave no hint of the burning ambition that consumed him. It is impossible to know how long he would have continued to quietly suffer Mazarin's manipulations had the cardinal lived. But at half past two in the morning of March 8, 1661, at the age of fifty-eight, Cardinal Mazarin passed away in his rooms at the château of Vincennes, comforted by his Italian confessor, Father Angelo. And for Louis XIV, overnight, everything changed.[12]

The Sun King

The cardinal's death had been expected for some time, and when Louis was informed later that morning, he was not surprised. He dressed himself and quietly went into his study, locking the door behind him. The next two hours, spent in seclusion, may have been the most consequential in the history of the monarchy. When he emerged he asked to see the department heads of the royal administration, and announced to them his intention to rule without a chief minister. "But who should we address ourselves to?" asked the ministers in confusion. "To me," answered the king, and so it was. For the rest of his long reign, every departmental decision and every official request had to be approved and signed by the king, in person.

To the twenty-three-year-old Louis it was as if a dam had burst, as all his pride and ambition, nurtured secretly for years, poured forth into the open. In the new regime the king, instead of serving as a figurehead for ambitious underlings, was the unchallenged and absolute ruler of the land, the source of all power, all legitimacy, and all favor. Finally giving vent to his simmering resentment of Mazarin, he denounced the very title of "chief minister" and declared that the office should forever be abolished in France. For, as he advised his son the dauphin some years later, there is "nothing more shameful than to see on the one hand all the functions and on the other the mere title of king."[13]

The sentiment was widespread, as Louis's assertion of personal rule was greeted with a wave of national euphoria that would carry on through the early years of his reign. It is not that Louis dispensed with the harsh methods practiced by his predecessors. Quite the contrary: when he thought it necessary he was just as willing as Richelieu and Mazarin to unleash the power of the state on any recalcitrant subject, the only difference being that he did so even more ruthlessly and efficiently. Local uprisings against intendants and tax collectors in the Boulonnais (1662), the Pyrenees (1664–1670), Bordeaux (1675), and Brittany (1675) were harshly suppressed, and hundreds of rebels and suspected rebels were sent to the gallows. Meanwhile the Parlements, whose pretensions to independent power had been a thorn in the side of the monarchy for generations, were quickly brought to heel and made into what the kings of France had always thought they should be: a rubber stamp that would automatically ratify all royal decrees. The message to the king's subjects rang loud and clear: no resistance to the sovereign's will, as represented by his decrees and his officials, would henceforth be tolerated.[14]

But if such measures had brought popular hatred and derision upon the cardinals, this was not at all the case for the brash young king. The two cardinal chief ministers, first Richelieu and then Mazarin, had served the monarchy well, but lacking the legitimacy of a sovereign they were cast as tyrannical usurpers. Louis, in contrast, was idolized. Some overtaxed peasants undoubtedly suffered, and noble parlementarians in Paris and the provinces undoubtedly grumbled, but throughout the kingdom Louis XIV's firm hand was greeted as a welcome change from the chaos of previous regimes. His brand of royal absolutism, which acknowledged no source of power and legitimacy other than the king, wasn't just accepted by his subjects as a price worth paying for peace and stability; it was, rather, enthusiastically embraced as the only proper and legitimate order of the state, and the path to national glory.

And little wonder, for the early years of Louis XIV's reign were indeed happy, even glorious. The newfound power and credibility of the state enabled Jean-Baptiste Colbert, Louis's *contrôleur général*, to overhaul the tax system and vastly increase the royal revenue. This did wonders for the health of the royal treasury, which in turn enabled the king to spend lavishly on his favored projects. First came the army, which was transformed from a heterogeneous collection of privately raised companies around a core of royal troops into a coherent, disciplined, and well-equipped body that even in times of peace numbered 100,000 men. Aided by a school for professional military engineers that refined its artillery, tactics, and fortifications, it quickly became the finest fighting force in Europe. Then came the navy, a military arm in which France had not traditionally excelled. Thanks to Colbert's personal patronage the vastly expanded and improved fleet soon equaled, and even surpassed, those of Holland and England, the traditional maritime powers. Seeking martial glory, Louis made immediate use of these powerful instruments. Between 1661 and 1672 he launched a series of aggressive and victorious wars that netted a significant expansion of the territory of France.[15]

Nor was Louis's determination to make his reign a glorious one limited to foreign adventures. Believing that the king's palaces must reflect his own brilliance, he launched a building campaign to renovate and expand the châteaus of Fontainebleau, Vincennes, Chambord, and Saint-Germain-en-Laye. Special attention was paid to the Louvre, the most visible royal residence, located as it was in the heart of Paris. The job of transforming the aging Louvre into a palace worthy of the great king was handed over to Claude Perrault, the man most responsible for the way the palace appears today. Splendid public entertainments held equal place in Louis's glorious vision, and in 1662 he organized a competition of knightly equestrian skill known as a *carrousel* in a square opposite the Tuileries. Five teams of nobles, fantastically dressed up as Romans, Turks, Persians, Indians, and Americans, competed for prizes, with Louis himself appearing as the "Emperor of the Romans" with a sun emblem on his shield. Not coincidentally it was around that time that he began referring to himself as the Sun King.[16]

Not content with brilliant display, Louis set about making Paris into the cultural and intellectual capital of Europe. He patronized the Académie Française and inducted the most brilliant men of letters in France into its ranks, including the playwright Jean Racine (1639–1699), the fabulist Jean de La Fontaine (1621–1695), and the fairy-tale author and collector Charles Perrault (1628–1703). He founded the Royal Academy of Sciences and soon counted the leading men of science in Europe (women were excluded) among its members, including the illustrious Dutchman Christiaan Huygens (1629–1695). It soon became, along with the Royal Society of London, the leading scientific institution in Europe. Alongside the Academy, Louis founded and built the royal observatory in Paris, the most advanced of its day, while along the Seine the modern new Collège Mazarin, founded through the late cardinal's bequest, was taking shape. In those early years of his reign Louis forever altered the skyline of Paris with the famous classical dome of the Invalides, a home and hospital for war veterans but also a monument to military glory in the heart of the city.

It was a remarkable transformation from the days of the regency, and it took place as if overnight. To contemporaries the change seemed miraculous. "During the regency," wrote Paul Pellisson (1624–1693), a leading historian and man of letters, in 1671, "you have seen impiety callously manifest; today it is either dead or mute at the court." In a previous reign, he continued, one could see "the license of the base-born corrupting persons of quality," but such outrages have now disappeared. And "in what part of the world was there formerly more laxity and indiscipline among individuals? In what part of the world today is it more difficult for men to avoid their responsibilities, abuse their authority . . . or neglect their duty?" All of this, concluded Pellisson, is due to a monarch who, from the day he ascended the throne, managed to "correct, overcome, and improve the customs, propensities, and genius of his people."[17]

Pellisson, it should be said, was a courtier and intimate of Louis XIV, so it is hardly surprising to find him singing the king's praises. Yet there is no doubt that he spoke for many in the first decades of Louis's reign who believed that they were living in a golden age, and that their young king had discovered the secret to peace, prosperity, and national glory. Louis, for that matter, believed so himself: his long humiliation at the hands of Mazarin and the unbearable outrages of the Fronde had led him to definite conclusions about the proper place and conduct of a king. Now that he had not only the title but also the powers of a king, he could finally act on what he had learned. As one success followed another, Louis was not surprised: it was merely the outward vindication of his vision of the monarchy and the state that he already knew to be true.

Louis XIV was not an intellectual, and it likely never occurred to him to compose a formal treatise on the theory of monarchy. He was, however, intelligent, strong-willed, and intensely conscious of his own place in history. Consequently he made a concerted effort for the benefit of future generations to write down his views on the politics of the day and the principles that guided his actions. Compiled and published decades after his death, Louis's notes add up to a very uneven history, covering certain years but not others, edited by a string of different courtiers, and ending in 1668—after which he recorded only his military campaigns. Yet despite its limitations, the *Mémoires for the Instruction of the Dauphin* is a remarkable document, offering a glimpse into a mind that in precisely those years reshaped not only France, but much of Europe. The distinctive voice of Louis XIV, proud and unreflective, yet well-meaning and—to his own mind at least—conscientious, resonates through the centuries on the pages of the *Mémoires*.[18]

To begin with, Louis explains, the world of men is not arbitrary, but organized according to what he calls a "natural legitimate order."[19] Accordingly, every person has his place in a well-defined social group, which is set in a precise relation to every other one. Each of these groups, furthermore, plays a critical role in the well-being of the entire social body:

> The peasant, by his work furnishes nourishment to this whole great body, the artisan by his craft provides everything for the convenience of the public, and the merchant by his cares assembles from a thousand different places all the useful and pleasant products of the world in order to furnish them to each

> individual whenever he needs them; the financier by collecting the private money helps to support the state, the judges by enforcing the law maintain security among men, and the clergy, by instructing the people in religion acquire the blessings of heaven and preserve peace on Earth.[20]

All these constituent groups are essential to the functioning of the whole and should therefore be honored and respected: "[far] from scorning any of these conditions," Louis writes, "we must take care to make them all, if possible, exactly what they should be."[21]

Respect, however, does not mean power. When Louis writes that the members of every station are to be made "exactly what they should be," he means that they should know their place and never challenge the decisions of their betters. "[Y]our maxim," he advises the dauphin, "must always be, like mine, to establish as far as possible the authority of those in command against those who strive to . . . escape from their control." Hierarchy must always be preserved, since "the tranquility of subjects lies only in obedience." Any effort on their part to assert political power is an offense to the natural order that is bound to lead to chaos, for "there is always greater evil in popular control than in enduring even the bad rule of kings." To conclude, "[i]t is perverting the order of things to attribute decisions to the subjects and deference to the sovereign."[22]

At the top of this immutable hierarchy, according to Louis, stand the sovereigns. They are the ones to whom all people owe obedience, whereas the sovereigns themselves owe obedience only to God. Such a relationship between people and ruler is bound to strike us as irredeemably oppressive, but to Louis it was natural: the people, he instructs the dauphin, "are almost part of ourselves, since we are the head of the body and they are its members."[23] And it is precisely because kings owe nothing to any man, and because they are the natural heads of the social body, that they are also the best rulers: for among all men "[t]he prince alone . . . has no fortune to establish but that of the state, no acquisition to make except for the monarchy, no authority to strengthen but that of the laws, no debts to pay except the public ones, no friends to enrich save his people."[24]

To us, the notion that the best ruler is one who owes nothing to his people is strikingly incongruous. Our democratic institutions are based on the opposite proposition, that those in power owe everything to the people, and we are happy to remind them, and quickly replace them, if they forget. But Louis thought otherwise: "Princes," he wrote, benefit from "brilliant birth and a proper upbringing," which "produce only noble and magnanimous sentiments." In contrast, he continued, assemblies of nobles or commoners derive all their decisions from base principles of utility. "The many heads who make up these bodies have no heart that can be stirred by the fire of beautiful passions."[25]

Could Louis really believe that commoners are slaves to narrow self-interest, and only princes are moved "by the fire of beautiful passions"? The obvious answer to anyone reading the *Mémoires* is that he believed precisely that, and there is no denying that he presented a coherent and self-consistent theory to that effect. Everything in the world, according to the *Mémoires*, is arranged in accordance with a fixed and unalterable natural order, and all men consequently live in a fixed and preordained relationship to one another. This order, for Louis, is functional, because it ensures that each social grouping serves society as a whole. And it is also hierarchical, because each grouping is assigned ranking in the overall scheme, a ranking that it must accept and never exceed. Inevitably, such a hierarchy leads to the one person at its apex, one who is supreme to all, and subservient to none—the king himself.

The sovereign, then, is not someone who imposes his will on his subjects by threat and by force. He is, rather, the necessary expression of an unalterable universal order. Any attempt to challenge his supremacy is not a matter of pitting one power against another, but a rebellion against nature. Absolute royal supremacy, for Louis, was not a matter of utility, or power, or one possible political system among others: it was the one true natural political order, and the only one that would bring about peace and prosperity.

It was this vision of the world, the state, society, and most of all himself that Louis so vigorously imprinted on his kingdom in those early years of his personal reign. And as victorious war followed victorious war, domestic peace prevailed, rebellions were crushed, and the glory of the Sun King shone from one end of Europe to the other, his vision was vindicated. The dramatic transition from the turbulent age of the cardinals to the glory of the king's personal rule was clear evidence that absolute royal supremacy was the only viable political system, and the only one that matched the natural order of the world.

The Geometrical State

The political order that Louis XIV was imposing on his kingdom in the 1660s and '70s with such remarkable success was a rational order: each and every person was assigned a role that was essential for the functioning of the body politic as a whole. The peasant, the artisan, the merchant, the lawyer, the nobleman, and—ultimately—the king all played their necessary parts in a mutually supportive society and state. It was also a strictly hierarchical order, rigid and unalterable, because one's position in the whole could never be challenged. It was universal, because, being the only legitimate political order, its principles were valid everywhere and always, and it was natural, because it was an expression of the deep order of the universe. In the seventeenth century this rational universal order that hierarchically structured everything in the world could only mean one thing: it was, inevitably, a geometrical order.

Indeed, as understood in the seventeenth century, geometry matched Louis XIV's vision of political order point for point. Just as Louis's state was rational and hierarchical, so was geometry, whose proofs followed a strict hierarchical sequence, beginning with self-evident assumptions and proceeding step by logical step to absolute, unchallengeable truths. Just as the absolutist state is rigid and unalterable, so is geometry, because its objects—triangles, squares, circles—are placed in a fixed and immutable relationship to one another. And just as, for the Sun King, absolute monarchy was universally applicable, so it was with geometry, whose truths apply everywhere and always. Finally, both the absolute political order of Louis XIV and geometry were viewed as "natural." Geometry was natural because, as Brunelleschi and Alberti had discovered, and Galileo proclaimed, the entire world and even space itself was geometrical through and through. The political order in turn was natural because, as Louis established in the *Mémoires*, it followed the natural order of the world.[26]

The connection between political absolutism and geometry seemed obvious to contemporaries, who frequently made use of geometrical metaphors to describe kingly rule. One was Cardin Le Bret (1558–1655), a jurist who had been one of Richelieu's chief advisers and devoted much time to justifying the Bourbons' centralizing policies. "Sovereignty," he wrote in his political apology for absolutism entitled *De la souveraineté du roi*, "is no more divisible than a geometrical point."[27] As any educated European of the age well knew, Euclid's definition of a geometrical point was precisely "that which has no parts." A chronicler as far back as 1649 referred to the king as "this center to which all lines from the circumference point," while Claude Nivelon, disciple and biographer of Louis's favorite painter, Charles Le Brun, similarly identified the king with the center of a circle: he sends orders "from the center of the state to its circumference."[28] And for the playwright Jean Racine the king illuminates the entire world "while scarcely moving from the center of his kingdom."[29]

The most systematic thinker in this vein was undoubtedly the Englishman Thomas Hobbes (1588–1679), who provided a complete geometrical theory of absolutism. Hobbes was no stranger to France: in 1610 he visited the kingdom shortly after the assassination of Henry IV and was profoundly affected by the chaos that followed the sudden removal of a monarch. In 1642 he once again found himself in Paris, this time as a refugee from troubles at home, where Parliament's challenge to King Charles I had led to outright civil war, and ultimately the king's execution. He remained in France for a full decade, much of it as mathematical tutor to the future Charles II.

Hobbes needed no convincing of the dire consequences that accrue to a kingdom that overthrows its sovereign. But in his philosophical writings he attempted not only to persuade his readers that this was the case, but to actually prove it, after the fashion of geometers. He therefore modeled his philosophical writings, and in particular his masterpiece, *Leviathan* (1651), on Euclid's *Elements*: starting with simple, self-evident assumptions, he argued slowly and systematically, step by logical step, until he reached a definite and unchallengeable conclusion. In the end, Hobbes proved that unless men entrusted their will to an absolute sovereign, they would be eternally doomed to a war of all against all and a life that was "nasty, brutish, and short." And he did so, to his own satisfaction at least, with absolute geometrical certainty.[30]

And it wasn't just that, according to Hobbes, an absolutist state was a geometrically proved necessity. It was also that the decrees of the absolute sovereign, the Leviathan, possessed the irresistible force of geometrical demonstrations. Only a ruler with such absolute power as to make opposition not only inadvisable but inconceivable, he reasoned, could protect the state from the ever-present danger of civil war. In Hobbes's geometrical state, resistance to the ruler was not only futile, but unthinkable.

Hobbes completed the *Leviathan* almost a decade before Louis XIV began his personal rule, but he lived long enough to see the Sun King in his full glory. And while many, and likely most, in England viewed Louis's pretensions to absolute power with apprehension, Hobbes was full of admiration. The French, for their part, returned the compliment. One of Hobbes's closest friends, as well as his French translator, was Samuel Sorbière, Louis's official "royal historiographer." To Sorbière, Hobbes was a "gallant man" and a true friend of "crowned heads," unlike other members of the English elite, whom he suspected of harboring republican sympathies.[31]

While Hobbes may have been sympathetic to the pretensions of royal absolutism, his Leviathan state was also deeply problematic for a king such as Louis XIV. For Louis, in the final account, was a king by divine right, raised by God Himself to his exalted position and charged by Him with the heavy responsibility of kingship. The Leviathan, in contrast, could be any man, as long as he was chosen by the people and entrusted with their free will. For Hobbes, the choice was a practical one, and could in principle be withdrawn if the Leviathan failed to live up to the people's expectations. This, needless to say, was as unacceptable to Louis as it was to Hobbes's own monarch, Charles II of England. The absolute geometrical hierarchy of the Leviathan state was undoubtedly appealing to kings; its populist foundations, however, seemed like an invitation to the overthrow of crowned heads.

The task of reconciling Hobbes's Leviathan state with the French monarchs' claims to divine election fell to Bishop Jacques-Bénigne Bossuet (1627–1704), a figure far more influential than Hobbes's friend Sorbière. For Bossuet was not only a leading intellectual in Louis's court, but also the king's spiritual guide and chief apologist for his brand of unfettered absolutism. His most famous work, *Politics Drawn from Holy Scripture*, more commonly known as the *Politique*, draws a great deal from *Leviathan*, even as it differs from it in crucial ways.[32] Like Hobbes, Bossuet argued that, if left to their own devices, men were doomed to a miserable existence of endless, pointless strife. But whereas Hobbes contended that reason would lead men to entrust all their will and power to the Leviathan, Bossuet's doctrine was more fitting for a monarch who styled himself the "Most Christian King": not reason, Bossuet argued, but only God could lift men from their fallen state, and He did so by appointing a king to rule over them. The king, it follows, owes nothing to the people and everything to God. The people, in turn, owe everything to the king, who, as God's elect, is more than a man: "You are gods," Bossuet declared to Louis in a sermon shortly after he had assumed personal rule. "You are gods, though you will die . . . The man dies, it is true, but the king, we say, lives on."[33]

Bossuet found much to criticize in Hobbes. The Englishman's contention that the ruler's power was ultimately derived from the will of his subjects was, of course, anathema to a legitimist monarchist. So was the fact that Hobbes's materialist reasoning did not allow for the operation of divine grace and left the world godless. Indeed, persistent rumors that Hobbes himself was an atheist did nothing to assuage the Frenchman's fears. Yet Bossuet fully agreed with Hobbes on his central point: that the only way to prevent men from descending into endless and vicious civil wars was through the rule of an absolute, all-powerful sovereign. And even more fundamentally, the English philosopher and the French bishop agreed on this: that the only proper way to make their argument was through strict and rigorous geometrical reasoning.[34]

For, remarkably, despite his mystical belief in the divinity of monarchs, Bossuet structured the *Politique* in much the same way as Hobbes had structured the *Leviathan*—as a rigorous geometrical treatise, modeled on Euclid's *Elements*. Accordingly, each of the ten books that compose the *Politique* is divided into "articles," which are general theorems, such as "Man is made to live in society" (book 1, article 1), or "Royal authority is subject to reason" (book 5, article 1). Each of these in turn is then demonstrated by a succession of propositions that are themselves also rigorously "proved." But whereas Euclid had relied on strict reasoning for his demonstrations, and Hobbes maintained that his systematic reasoning was as valid as Euclid's, Bossuet relied on something else: quotations from scripture. Each proposition is proved by reference to a particular passage in the Bible, along with an explanation of how the quotation bears on the question at hand.

It was a remarkable juxtaposition of Biblical literalism and the rigorous rational structure of geometry. Yet to Bossuet, Louis, and many of their contemporaries it made perfect sense. They never questioned that kings were God's elect, and that scripture held the key to the mystery of their power. Yet at the same time they also knew that kingly power, as manifested in the world, was rational, hierarchical, and rigorous, or in other words, geometrical. Bossuet's unusual treatise combined both ends of kingship—its mystical roots and its geometrical expression.

Indeed, almost everything in the court of Louis XIV was arranged with the rigidity and precision befitting a geometrical demonstration. First, and most critical, was the strict hierarchy that governed all goings-on at court. At the top was the king, of course, followed by the dauphin and his wife, monsieur (the king's brother) and madame (monsieur's wife), and the children and grandchildren of the dauphin, all known as the Children of France. Next came the "Princes of the Blood," the cardinals, the ecclesiastical peers of France, the dukes who were the lay peers of France, the "life dukes" who were not peers, and finally the lowest and most numerous class, consisting of mere counts, marquises, and barons. To these should be added the foreign princes residing in the court, whose ambiguous positions were the cause of endless bickering.

The hierarchies in Louis XIV's court were complex, elaborate, and refined, and they were no laughing matter. At the king's supper, who should be seated closer to the king? If two nobles meet, who could sit in whose presence, and who must stand? And if allowed to sit, sit in what? Who should be allowed an armchair, and who relegated to an armless chair, or even a three-legged stool? And if sorting this out was not challenging enough, the complexity level grew exponentially whenever a larger group came together. How could one sort out the standing of each person in relation to all others? The questions were not trivial, and errors had consequences.

Mme de Torcy, a minster's wife who by accident or design took a place at supper above that of a duchess, brought upon herself a storm of indignation from the outraged king. "Madame," married to the king's brother, wished to visit her daughter, the Duchess of Lorraine, and as was the custom, sent out emissaries to negotiate the order of precedence. But when the daughter's husband, the Duke of Lorraine, insisted on his right to sit in an armchair in Madame's presence, the visit was called off. The fifteen-year-old Duchess of Berry, a granddaughter of France, was shocked when the attendant usher opened the two leaves of a double door to allow her mother to enter her rooms. The mother, a mere princess of the blood, was entitled to have only one leaf open when entering the presence of a senior in rank. When her demand to immediately dismiss the usher failed (he was the king's servant, not her own), the duchess "weeps and storms."[35] Sometimes the hierarchies followed explicitly geometrical patterns: when the Parlement of Paris opened its sessions, it did so in the presence of the great nobles of the realm. But whereas princes of the blood could proceed to their seats directly, diagonally crossing the floor of the "Grande Chambre," this was strictly forbidden to dukes. They were required to walk along the walls, turning the corners at right angles to reach their benches.[36]

Such was life in the court of Louis XIV, where hierarchy determined all and performing it correctly was the substance of each and every day. To the courtiers this was not a matter of choice: the king insisted on it, and he himself set the highest standard in the correct performance of his own exalted position. Even the king's critics conceded that whatever one thought of his policies, no king ever looked the part more than Louis XIV. This was no trivial matter: "Those who imagine that claims [of precedence] of this kind are only questions of ceremony are sadly mistaken," Louis explained in the *Mémoires*. "There is nothing in this matter that is unimportant or inconsequential." The acting out of his own supremacy is one of the most important things a monarch does, since "one cannot deprive the head of state of the slightest marks of superiority without harming the entire body." Consequently, such public performance is a duty the king owes to his subjects, and "we should guard nothing more jealously than the preeminence that embellishes our post. Everything that indicates or preserves it must be infinitely precious to us."[37]

As a result, life in the court of Louis XIV was an endlessly complex dance in which gesture, attitude, and position depended on the situation and the persons present, and were readable only to those who spent their life deciphering the codes of courtly behavior. When taken as a whole, the court was a living demonstration of the world as Louis XIV saw it: perfectly ordered, rigidly hierarchical, and completely under the sway of the monarch, the apex of the pyramid, where all hierarchies end.

Distilled and purified into high art, the daily dance of life at court became in reality a new kind of dance—the ballet.[38] To be sure, ballet was not invented in the age of Louis XIV. We had already seen how as far back as 1573 Catherine de Medici's ladies in waiting dressed up as nymphs and danced a complex geometrical ballet in honor of the future Henry III, who had been offered the crown of Poland. Only eight years later Catherine's court dancing master, Balthasar de Beaujoyeulx (1535–1587), created an even more spectacular ballet for the wedding of King Henry's sister-in-law, Marguerite de Vaudémont. The *Ballet comique de la reine*, as it was called, set the standard for elegance and cultural refinement for decades to come.

Described by a contemporary as "a uniquely creative geometer," Beaujoyeulx proved worthy of the moniker. In the *Ballet comique* the dancers traced circles, squares, and triangles across the floor, performing the deep geometrical order that pervaded the universe. "So dexterously did each dancer keep her place and mark the cadence," Beaujoyeulx wrote, "that the beholders thought that Archimedes himself had not a better understanding of geometrical proportion." And, just like the elegant geometrical parterres of the Tuileries, the ballet presented the promise of a peaceful and prosperous kingdom, rationally ordered under the monarch's benevolent gaze. It was a promise that stood in sharp contrast to the realities of Henry III's war-torn kingdom, and Beaujoyeulx was not shy about stating his purpose: "[A]fter so many unsettling events," he wrote the king in the ballet's preface, "the ballet will stand as a mark of the strength and solidity of your Kingdom." The royal geometrical dance, like the royal geometrical gardens, had become a tool of state.[39]

It was not, however, in the beleaguered household of Catherine de Medici, but in the status-obsessed court of Louis XIV a century later that ballet was perfected into high art. Back in the troubled years of civil war the dance had stood for the distant promise of a harmonious kingdom founded on the geometrical order of the universe. But under the glorious reign of the Sun King that promise was allegedly fulfilled, and the court celebrated the new golden age with elegant ballets in which the highest nobles of the realm danced the key parts. Among the most ardent performers was Louis himself, an enthusiastic and accomplished dancer who participated in no less than forty major ballets as a young man, and followed others closely into old age.

With the king leading the way, the ballets, known as *la belle danse*, became central to life at court, and dancing an indispensable skill for the aspiring courtier. The elegant movements of the ballet were no longer just representations of harmonious geometrical order, but also ritualized performances of the hierarchies that governed court life and the kingdom as a whole. From the great dancing masters, such as Raoul-Auger Feuillet (1659–1710) and Pierre Rameau (1674–1748), one learned how to perform the elaborate etiquette of the court with grace and an air of natural ease. The refined art of how to bow, how deeply, and to whom; how to greet an inferior or take off one's hat to a superior; and how to enter and exit a room with unforced elegance was indispensable for survival in the elaborately stratified court of Louis XIV. Gracefulness or awkwardness in the performance of such routine encounters could prove the difference between brilliant success and miserable failure.

The geometrical rhythms of *la belle danse* formalized these daily rituals of court life and distilled them to their essence. Like the grand performances at the court of Catherine de Medici, ballet in the age of Louis XIV evoked the geometrical harmonies that govern the world. But unlike a century before, this deep order was now hierarchical through and through. The movements of the dancers and their interactions with one another and with the audience were emblems of courtliness, a precise alternating of deference and assertiveness, all performed with an air of natural ease. That the movements—though taught, studied, and endlessly practiced—appear effortless was as crucial for daily life at court as it was in an elaborate and scripted ballet. Any awkwardness or artificiality of movement would inevitably cast doubt on the deep truth that both dance and court etiquette were intended to convey: that the rigid and regimented order of Louis XIV's court and kingdom was simply the natural order of the world.

And, naturally, this order was geometrical. The postures of the dancers, first codified by Louis's personal ballet master, Pierre Beauchamps (1631–1705), were based on exactly five positions, each defined through the precise alignment of the limbs and the prescribed angle between feet, legs, and arms. When followed correctly, they prepared the body to move in straight lines and right angles—forward, backward, or sideways.[40] In the ballets themselves the dancers traced symmetrical figures—loops, circles, and S-shapes. Performing sometimes solo and sometimes in couples, the dancers moved synchronously and mirrored each other around the axis of the seated king, in ways that could not but recall the elegant geometrical parterres of the royal gardens, reflecting each other across a central path descending from the royal palace. When at the king's request Beauchamps, and later Feuillet and Rameau, developed a special dance notation, they were, in essence, reducing the courtly dance of universal harmony and social hierarchy to a geometrical diagram. Some decades later, starting in 1751, Denis Diderot and Jean Le Rond d'Alembert published the *Encyclopédie*, often described as the greatest publishing venture of the age of Enlightenment. It was only appropriate that when they came to the entry discussing dance notation systems they assigned it to Louis-Jacques Goussier, an illustrator, engineer, and, most important, a mathematician.[41]

In the kingdom of Louis XIV geometry reigned supreme. Court rituals were designed as precise geometrical sequences, and courtly etiquette was a carefully choreographed geometrical dance. Ballet, which elevated courtly grace to a high art, was founded on geometrical principles, as was the philosophy of royal absolutism, which was modeled on Euclid's *Elements.* Bernard Le Bovier de Fontenelle, the perpetual secretary of Louis's Academy of Sciences, gave voice to what to many was an unquestioned assumption when he suggested that pretty much any field of endeavor should be structured geometrically. "A work on ethics, politics, criticism, and, perhaps, even rhetoric," he wrote in 1699, "will be better . . . if done by a geometer."[42]

Yet none of these fields could compare with the ultimate expression of the geometrically ordered land: the palace and gardens of Versailles. Although they are forever associated with the outsize ambition and magnificence of the Sun King, the original château and gardens of Versailles were, in fact, created by his father. In 1624 Louis XIII had a lodge built at the site, to be used on his hunting expeditions in the neighboring woods. The king, it seems, thoroughly enjoyed his visits, because seven years later he decided to upgrade his residence, replacing the wooden structure with a sizable château and gardens. By the time the future Louis XIV was born in 1638, Versailles was one of the regular stops on the annual tour of the court through the royal palaces of the Île-de-France.[43] The Dauphin, and later the young king, grew to love the quiet bucolic setting that afforded him a welcome respite from the pressures and indignities of Paris. And so it was that one of Louis's first actions in 1661, when he became king in fact rather than just in name, was to order the massive expansion of the gardens of his favorite château at Versailles.

It is very likely that by the time of Louis's visit with the unfortunate Fouquet on August 17, work on the new design of Versailles was already under way. Yet what the king experienced at Vaux-le-Vicomte that day made it clear to him that another garden in the style of the Tuileries or Luxembourg simply would not do. To eclipse the brilliance of his errant finance minister, he would have to turn to the very team that had created for Fouquet the most beautiful estate in France: the painter and designer Charles Le Brun, the architect Louis Le Vau, and—most important—the gardener André Le Nôtre.

The Man Who Knew His Place

To say that André Le Nôtre was born to be a gardener is not metaphorical praise of his genius but a statement of fact. André's grandfather was Pierre Le Nôtre, master gardener to Catherine de Medici, and his father was Jean Le Nôtre, who bore the title of First Gardener to the King. As befits a clan of royal gardeners, the family lived and worked in the Tuileries, and that was where the future designer of Vaux-le-Vicomte and Versailles was born in 1613. Scion to a gardening dynasty, born on the grounds of the premier royal garden in France: there was never any doubt what vocation young André was expected to pursue. Some youths might perhaps rebel against their preordained destiny, but that was not Le Nôtre's way. He accepted his vocation as his allotted place in life and followed meticulously the path laid before him. "Never did he overstep his position" was the high praise accorded to Le Nôtre many years later by the courtier and diarist the Duke of Saint-Simon (1675–1755), who knew Le Nôtre toward the end of the latter's life. And that indeed is an apt description of the man: For Le Nôtre, even from his earliest days, was a man who knew his place.[44]

It worked well for him. Growing up he undoubtedly learned the family trade from his elders, so that as a young man he already knew how to plant and maintain the great rectangular parterres that were the building blocks of the great French royal gardens. Practical training, however, was not enough: it was in those very years that Jacques Boyceau was elevating the art of gardening from a form of manual labor into a theoretical discipline worthy of a gentleman like himself. Echoing Alberti, who insisted that an architect could not do without "Painting and Mathematicks," Boyceau in his *Traité du Jardinage* explained that a true gardener must be a student of painting, architecture, and geometry. As Louis XIII's "*Intendant des jardins*" Boyceau was the immediate superior of Jean Le Nôtre, who wisely took the advice to heart. He accordingly dispatched young André to expand his education in the studio of the royal painter Simon Vouet (1590–1649), where he studied draftsmanship, the principles of geometry, and the theory of linear perspective. As Boyceau had foreseen, these skills would prove indispensable for Le Nôtre as a high-level gardener in the royal service.

For that is indeed what Le Nôtre was soon to become: in 1635, at the age of twenty-two, he was already chief gardener to the king's brother, Gaston d'Orleans, with special responsibilities in the Luxembourg gardens. By 1643 he was appointed *dessinateur de parterres de tous les jardins du roi*, and in 1649 he inherited his father's former position as *premier jardinier du roi* at the Tuileries. Finally in 1657 he became *contrôleur général des bâtiments du roi*, a position that gave him responsibility for the design and maintenance of all royal residences. For a man born to be a gardener, it was a meteoric rise. Even before the first tree was planted or the first parterre laid out in the garden that would immortalize him, Le Nôtre had already reached the highest position of any gardener in the annals of the royal service.

How did a man born in a humble house on the grounds of the Tuileries become a high official in the royal court, and ultimately a personal friend and confidant of the king himself? His personal charm and warmth, which was widely attested to by contemporaries, probably had something to do with it. An anecdote reported by his nephew Claude Desgots tells how during an audience with the pope in Rome in 1679 he became so engaged in conversation that he forgot himself, slapped the aged pontiff on the shoulder, and then hugged him. When Le Nôtre reported the tale in a letter to Versailles, one of the king's courtiers offered to wager that even the notoriously informal gardener did not dare to embrace the pope. "Do not bet on it," the king responded. "When I return from a campaign Le Nôtre always embraces me. It is very possible that he embraced the pope."[45]

Partly it was also due to the fact, noted by Saint-Simon, that for all his success Le Nôtre was a man who knew his place. Not only did he not seek out honors that exceeded his social station, he carefully deflected them when they came his way. When in the 1690s, after decades of work on Versailles, the grateful king offered to grant Le Nôtre a proper coat of arms, the gardener turned it into a joke. He already had a coat of arms, he replied, consisting of three snails surmounted by a cabbage. "Sire, how could I forget my spade?" he exclaimed, gently reminding Louis of his origins as a simple laborer. "Is it not to that I owe the bounty with which Your Majesty honors me?" In this instance, at least, the king was not deterred, and bestowed the coat of arms anyway: it consisted of a gold chevron and three silver snails.[46]

Le Nôtre was, as we have noted, born to be a gardener, but he was also, by nature and temperament, born to be a courtier for Louis XIV. His charm and bonhomie were incalculable assets in a world in which social interaction formed the essence of each and every day, so much so that even his transgressions of the strict court etiquette served to enhance his stature and endear him to his betters. The reason was that to all who knew him Le Nôtre was, self-evidently, precisely the kind of man that Louis XIV's court was meant to foster: a man who embraced his allotted station in society and the profession he was destined for from birth, and whose passion in life was to perform to the utmost of his abilities the role that fate had assigned him. For such a man, even the occasional inadvertent transgression only reinforced the fact that for him, playing out his precise role in a hierarchical society was not a forced exercise, but was as natural as breathing. In his passion to perform his station in life, Le Nôtre was the equal of his sovereign, Louis XIV; he was also the opposite of Mazarin's superintendent of finances, Nicolas Fouquet, who, though a commoner like himself, most emphatically did not know his place. In Louis XIV's France the unassuming gardener thrived, while the magnificent financier spent the last decades of his life in a lonely prison fort.

The Garden of Ambition

If Le Nôtre was the ideal courtier, it was not only because he performed the role of humble gardener to perfection. It was also because his highest aspiration, and his life's goal, was to be the very best at his profession that he could possibly be. In Le Nôtre's case this meant being the leading gardener in France, and to future generations perhaps the greatest gardener that ever lived. The chance to demonstrate his genius came his way in 1656, when Fouquet hired his services and charged him with creating the dream estate of Vaux-le-Vicomte. For a man in the king's service this was a rare opportunity to design a garden from scratch and supervise every step of its transformation from a paper plan to living reality. Most royal gardens, such as the Luxembourg or the Tuileries, had been in place for decades, even centuries, and all a landscaper could hope for was to gently improve on an existing pattern. Vaux, by contrast, was a blank slate on which an entirely new pattern could be created. Over the previous decades Fouquet had purchased the land surrounding the original estate, removed two hamlets and the village of Vaux-le-Vicomte, and leveled the ground, creating a gently sloping terrain from north to south. It was the perfect canvas for a gardener trained as a painter to showcase his skill.[47]

We have already seen the garden Le Nôtre created for Fouquet at Vaux-le-Vicomte. As at Versailles some years later, the château was approached by three arrow-straight boulevards that converged at the entrance.[48] The garden, mostly hidden from the front, spread out to the distance behind the château when viewed from the ground floor. A clear central axis ran straight down from the middle of the château all the way to a statue of Hercules visible on the horizon. Two perpendicular axes, marked by straight canals, bisected the central axis at right angles, completing the geometrical layout of the gardens. Elaborate parterres and elegant pools lined the central path symmetrically on both sides, and the intersections with the perpendicular axes were marked by fountains, tranquil pools, water cascades, and a grotto.

None of the elements Le Nôtre created at Vaux were in themselves new. Fountains, pools, and elaborate parterres had been the key components of French royal gardens for more than a century. Yet no one visiting Fouquet's dream house, not least of all the king himself, could doubt that in its total effect Vaux was startlingly new. Gone were the rectangular sections so familiar from the Tuileries, ranged next to one

FIGURE 17: **The Gardens of Vaux-le-Vicomte today**

another in monotonous rows, like rooms in a house. Gone were the gradated terraces of Saint-Germain-en-Laye, arranged one on top of the other as a succession of distinct gardens. And gone were the oddly shaped contours of the Luxembourg gardens, where an impressive, centrally located "grand parterre" led nowhere, with the main axis of the garden trailing off at an odd angle.

Vaux-le-Vicomte was something else: a carefully balanced and integrated whole, stretching from the palace to the horizon along a clearly defined axis. Every element in the composition supported every other one, and each reflected on the central axis and pointed to the château at the top of the hill. In the Tuileries one could enter one of the rectangular segments and pay no attention to the others; in Saint-Germain-en-Laye one could wander a terrace and ignore the others; at the Luxmbourg one could trail off to the end of the gardens and effectively lose sight of the grand parterre and the château. But not so at Vaux. No parterre, pool, or fountain could exist as it did without reference to all the others, to the overall geometrical arrangement and symmetry, and to the château that presided over it all. The whole depended on its parts, and the parts on the whole, just as in the composition of one of the paintings Le Nôtre had studied under Vouet's tutelage.

Vaux was indeed a painting in flowers, greenery, water, and stone, as carefully crafted as a composition by one of the Italian masters. And just like a Renaissance painting, it was ruled by the principles of linear perspective. Brunelleschi, Alberti, and Masaccio had shown how to create a geometrically determined three-dimensional space within their paintings by establishing a "vanishing point" where all parallel lines converge. To re-create the effect in a garden, André Mollet (ca. 1600–1665), who like Le Nôtre was scion to a dynasty of royal gardeners, suggested posting an actual painting at the far end of a garden to complete the view.[49] But Le Nôtre did not require such artificial embellishments: the statue of Hercules, situated on the horizon at the far end of the central axis, was effectively the vanishing point of the "painting" of Vaux-le-Vicomte. When viewed from the ground floor of the château, all the parallel lines outlining the parterres and pools converged on this point, turning the elegant gardens into a perfect perspectival painting.

At Vaux, Le Nôtre had re-created in three-dimensional reality the geometrical space that the master painters of the Renaissance had depicted on their two-dimensional canvases. It was a world in which every point was geometrically determined by the principles of linear perspective, and, as in a well-composed painting, every object related to every other and could not be discarded without destroying the harmony of the whole. It was a world in which everything had its uniquely defined and unalterable place in a grand hierarchical scheme, leading up to the château. It was also a world in which people would find and fill their assigned places under the benevolent gaze of the château's master. It was a peaceful, harmonious, geometrical world in which everyone knew their place, with a single exception: Fouquet himself.

When Louis XIV visited on that fateful summer day in 1661, he saw and noted it all: the geometrical order, the rigid hierarchy, the elegant unalterable harmonies, and the peaceful, prosperous, exquisite landscape that all of these engendered. It was a perfect world organized according to its "natural legitimate order," precisely as the king had presented it in his *Mémoires*.[50] It was, to Louis's eye, in all respects a royal garden, and superior to all others: its message of peace and harmony through hierarchy and order rang forth loudly and clearly, more so than even the Tuileries, Saint-Germain, or Luxembourg could impart.

Except, of course, that Vaux was not a royal garden, but a private one, belonging to a lowly minister. Except, of course, that the apex of Vaux's hierarchies was occupied not by the "natural legitimate" ruler, the king, but by a mere commoner, Nicolas Fouquet. Is it a wonder that the Sun King found such a situation outrageous beyond words? This was, after all, the Louis who insisted in his *Mémoires* that "one cannot deprive the head of state of the slightest marks of superiority without harming the entire body," and that "we should guard nothing more jealously than the preeminence that embellishes our post."[51] It was Louis who insisted that the enactment of precedence and order be carefully observed, and whose courtiers were engaged in a perpetual performance of their proper places in the hierarchy. And Louis who raged for hours after a minister's wife committed the unforgivable sin of taking a higher seat at table than a duchess. What, then, would Louis think of a commoner who not only exceeded his position, but also presented himself to all the world in the guise of a king? Such presumption had to be not only punished but crushed.

For Louis, Vaux-le-Vicomte was not simply an expression of vanity or personal arrogance on the part of an untactful minister. It was a direct attack on the monarchy, the state, and the king himself. The social order and the peace of the realm, he believed, depended on strictly preserving the elaborate social hierarchy, and that in turn depended on presenting it publicly for all to see. Most critically, the absolute supremacy of the king must never be in doubt. "Our subjects," Louis warned the Dauphin in his *Mémoires*, judge things by their appearances. It is therefore critical that the sovereign "be raised so far above the others that no one may be confused or compared with him."[52] Yet at Vaux-le-Vicomte Fouquet had done precisely that: with his magnificent geometrical gardens he had raised himself to the level of a king, thereby challenging the deepest foundations of Louis XIV's state. As Louis saw it, if such a challenge was allowed to stand, royal rule would be irreparably compromised and the kingdom would revert to the dark days of chaos and civil war.

To prevent this, Louis acted as he knew he must. He arrested the upstart superintendent, kept him a prisoner till his death, and stripped the offending estate of its treasures, sending them to adorn his own château at Versailles. The message was clear, and echoed throughout the kingdom: such magnificence as the errant minister had sought for himself belongs to no man but the king. Even this, however, was not enough: to demonstrate his incontestable supremacy over all his subjects, Louis also needed to prove that he could so far outshine Fouquet's reckless challenge that no one would dare repeat it. And so, along with the trees and statuary of Vaux, he also co-opted the men who made Fouquet's dream a reality—Le Vau, Le Brun, and, most important, Le Nôtre—and set them to work on his own dream palace at Versailles. There they would create for him a landscape so magical and wondrous that it would entirely eclipse the memory of the superintendent's folly. No one could ever outshine the glory of the Sun King, and Versailles would prove the point. It was Louis's unequivocal answer to Vaux-le-Vicomte.

The Enchanted Isle

The gardens that Le Nôtre created at Versailles were like no others in the world. Contemporaries called them "gardens of pleasure" to distinguish them from the utilitarian gardens that provided nourishment to castles and monasteries in the countryside. Indeed, with the exception of a small extension known as the *potager du roi* dedicated to the king's table, there was nothing "useful" about them.[53] Nothing found in the gardens—not the flowers, trees, parterres, fountains, canals, or bosquets—could help support the burgeoning population of courtiers in the palace and town of Versailles for even a single day. And yet "pleasure gardens" though they are, the gardens of Versailles are not as pleasurable as one might expect.

Impressive they undoubtedly are: their sheer size is awe-inspiring, their richness and variety is overwhelming, the meticulous geometrical design of everything from a single bush to the ground plan of the whole gardens is overpowering. But all that, as visitors even in our own day can attest, does not necessarily add up to an entirely enjoyable experience. Visitors to Vaux, then and now, take great pleasure in strolling down the broad central allée from the château, between the parterres, past the fountains, and down to the grand canal. Visitors to Versailles are more likely to experience sensory overload. They will be lost in the complex layout of the gardens, paralyzed by the conflicting pull of so many different parterres, fountains, and bosquets, challenged by the ubiquitous classical statues, each with its own story, and ultimately crushed by the sheer vastness of the estate. Dazed and confused, they will likely acknowledge defeat after visiting only a tiny fraction of the gardens, vowing to return another day to complete their tour. Most, needless to say, never will.[54]

The complaints are not new. Louis XIV's courtiers, to be sure, sang the gardens' praises, as one would expect from men and women whose livelihood depended on the king's pleasure. André Félibien (1619–1695) was Louis's court historian and one of the luminaries who had made the transition from Vaux to Versailles. Charged by the king with describing the gardens as they took shape and the celebrations that took place on their grounds in the early years, he produced a steady stream of elegantly written laudatory accounts. The novelist Madeleine de Scudéry (1607–1701), who had also been one of Fouquet's protégés at Vaux, authored a beautiful fictional account of a tour of the gardens.[55] Others were also impressed, including the English naturalist Martin Lister (1639–1712), who proclaimed after a visit that Versailles's splendor surpassed anything found in Italy, and the Russian ambassador, who declared the gardens worthy of the biblical King Solomon.[56]

Others were less complimentary. One of these was the Duke of Saint-Simon, who authored a detailed diary of life in Louis XIV's court, but prudently kept it to himself until decades after the king's death. "The gardens astonish by their magnificence, but cause regret by their bad taste," he pronounced, concluding that "[t]he violence everywhere done to nature repels and wearies us despite ourselves." His complaint that "[y]ou are introduced to the freshness of the shade only by a vast torrid zone, at the end of which there is nothing for you but to mount or descend" should be familiar to anyone who has visited Versailles on a hot summer day.[57] Another critic of the gardens was the Jesuit architectural theorist Marc-Antoine Laugier (1713–1769), who denounced the unmitigated monotony of the straight allées, the parterres, and the bosquets.[58]

It is doubtful, however, whether any of these laments would have troubled Louis XIV in the least. The purpose of the gardens, after all, was not to entertain: it was to impress, overwhelm, and overawe all comers with the glory and irresistible power of the Sun King. This was not always a pleasant message, or one that was necessarily welcome to those who wandered the garden paths. It is doubtful, for example, whether the Doge of Genoa much enjoyed his visit to the gardens in 1685, when he was summoned to Versailles after the French Fleet had pulverized his city. But he unquestionably got the message, and he made his obeisance to the king.[59] It was a message that Louis XIV wished to impress upon all his subjects, indeed all the world, and it was the message the gardens were designed to deliver. And deliver it they did.

André Le Nôtre arrived at Versailles around the fall of 1661 and immediately began drafting new plans for the gardens. Unlike Vaux, which was created from scratch, Versailles was already an established royal residence surrounded by a garden of considerable size, and in his earliest designs Le Nôtre tried to incorporate the existing layout into the one he saw in his mind's eye. It did not take long, however, for the courtly gardener to perceive that the king would never be satisfied with a mere improvement of an already-existing garden. What Louis required was a garden that would dwarf its predecessors in both scale and artistry, and forever outshine all others. To accomplish this, Le Nôtre expanded the garden far beyond its original borders to cover areas that had previously been forests and marshes. There, unencumbered by previous designs, he was free to shape the landscape as saw he fit, just as he had done at Vaux. It required three full decades of his undivided attention, but there is no question that Le Nôtre, in the end, succeeded: he completely transformed the original estate, leaving little to remind one of its original design and contours. In its place stood the brilliant new home and capital of the Sun King.

In his early years at Versailles, between 1661 and 1664, Le Nôtre focused on revamping the area close to the château, which soon came to be known as the Petit Parc, in contrast to the Grand Parc farther down the valley to the west. Although small in comparison to the eventual area of the estate, it was already a large enclosure, its roughly 180 acres three times the size of the Tuileries. Nor was it virgin land: a map of the gardens from shortly before Le Nôtre arrived on the scene clearly shows that the old gardens, dating from the reign of Louis XIII, covered much of the same area.[60] While Le Nôtre maintained the general layout of this early garden, he revamped and enhanced its basic geometrical design.[61]

The general shape of the gardens at this time was a symmetrical trapezoid. One side included the château itself and extended equal distances to the north and south of it; a parallel and slightly longer side was located about half a mile to the west, and two straight and equal sides joined the ends of the two parallels to complete the trapezoidal shape. Under Le Nôtre's supervision the main axis, which was ambiguous in the old design, was widened and enhanced, and made to bisect the trapezoid from east to west. Beginning as a broad avenue emanating from the precise center of the château, the axis passed between two symmetrical and elegant oblong-shaped parterres, each with colorful flowers arranged in elaborate and delicate patterns to look like an embroidered oriental carpet. Continuing past a circular pool, it descended in a monumental horseshoe staircase to a grand oval fountain, which within a few years was to become Latona's Fountain (Le Bassin de Latone), as it is still known today. Onward past two more elongated parterres adjoining the fountain in the same basin, the avenue continues westward, straight and broad and sloping gently downhill. It runs past large wooded bosquets, two on each side, before arriving at the largest fountain in the gardens. Known at the time as the Fountain of the Swans (Bassin des Cygnes), and soon to become the famed Fountain of Apollo (Bassin d'Apollon), it was elegantly shaped as superimposed ovals and a rectangle. This fountain lay at the center of the far side of the trapezoidal garden, and marked the end of the western progression along the central axis.

The monumental central axis, appropriately known as the Allée Royale, was flanked by two subordinate east–west axes, which ran parallel to the main one on both sides, each at about half the distance to the garden's northern and southern borders. Like the main axis, they consisted of arrow-straight avenues running from the eastern boundary of the garden, which was an extension of the château, to the western boundary, which ran through the center of the Fountain of the Swans. They were, however, significantly narrower than the central avenue and were interrupted on their course only by one single small pool each. The three east–west axes were, in turn, intersected at right angles by four north–south axes connecting the two angled sides of the trapezoid. These avenues were similar in size to the two secondary axes and ran mostly uninterrupted from one side of the park to the other. The intersection of the three parallel avenues with the four perpendicular ones created a grid pattern of fourteen roughly rectangular spaces to be filled however Le Nôtre and his assistants saw fit—with parterres, fountains, statuary, and bosquets. Indeed, much of the work of Le Nôtre and his successors at Versailles was devoted to perfecting these enclosures, turning them into self-contained worlds of magic and surprise.

Versailles in the 1660s was neither Louis's main home nor the seat of the court and administration, all of which remained in the Louvre for the time being. But the king grew increasingly enchanted with his country estate, so much so that in 1664, as Le Nôtre's first expansion of the gardens was drawing to a close, he invited the court to Versailles to share his enthusiasm. "The Pleasures of the Enchanted Isle" (*Les Plaisirs de l'isle enchantée*), as the celebrations were called, were five days of feasting and magnificence. The composer Jean-Baptiste Lully

(1632–1687) dazzled the six hundred assembled courtiers with ballets and musical entertainments, some familiar and some newly composed for the occasion. Molière was put in charge of the theatrical productions, which included among others *Les Fâcheux*, the very play he had produced for the king's visit to Vaux-le-Vicomte. Louis himself, radiant in the armor of a Greek warrior, shone in the jousting tournament and the horsemanship competitions, presenting his prizes alternately to the queen and to his mistress, Louise de la Vallière. Three monsters rose from the depth of the Fountain of the Swans, carrying nymphs who joined in a ballet, and at mealtime the illumination was so lavish it could have been daylight.[62] It was the king's unequivocal answer to Fouquet's grand entertainment that had so enraged him three years before. He considered it such a success that he repeated it four years later, this time condensing the nearly weeklong entertainment into a single brilliant day.

By this time, however, the Sun King was no longer content with the modest château inherited from his father, or even the magical and extensive gardens Le Nôtre had created for him. Versailles might be a sizable royal residence and a fitting site for lavish entertainments. But Louis was now considering moving his entire court there, and for that the existing château and gardens were simply inadequate. Colbert, who had inherited Fouquet's position as minister of finances and was the closest thing Louis had to a chief minister, had serious doubts about the proposed move, and was not shy about voicing them.[63] The constant work at Versailles was a drain on the royal finances, he argued, and it was also unwise to abandon Paris, the capital that Louis's grandfather had fought so hard to gain. "What a pity," he warned the king, "that the greatest and most virtuous king should be measured by the yardstick of Versailles!"[64]

The king, however, would not be dissuaded. In part this was because ever since the days of the Fronde, when he had found himself the prisoner of a restless populace, Louis had viewed the capital and its citizens with distrust. The move to Versailles would put a safe distance between the monarch and the unruly Parisian masses, and ensure that the city mob would never again trespass on his sacred person. This, however, was not the only or even the chief reason for the king's determination to move his court to Versailles. Far more clearly than the hardheaded Colbert, Louis understood the critical importance of royal display in an absolute monarchy. In Paris the king in his palace always had to contend with rivals: the Parlement, the church and its cathedrals, the wealthy merchants in their hotels, and most of all the bustling, restless city itself, which completely surrounded and largely overshadowed the royal palace. The king was but one voice in a cacophony, the palace but one seat of power among several. At Versailles all would be different. Here the sovereign would stand alone, his brilliance undimmed by other suns; here he could present himself in a state so exalted that no person or institution would dream of rivaling his glory.

And so in 1668 Louis embarked on a project that would transform Versailles from a delightful royal residence into the heart and soul of the kingdom, the shining beacon toward which all eyes turned. First, he asked his architect Le Vau to vastly expand the château, while preserving at its center the familiar structure that he had loved since childhood. Le Vau set to work, designing two new wings that hugged the original structure on the north and south and an elegant façade and terrace on the west side that overlooked the gardens. The new stone additions effectively encased the old brick château and became known, fittingly, as "the envelope." A decade later Le Vau's successor, Jules Hardouin-Mansart (1646–1708), began work on the massive long wings to the north and south of the main château, ultimately creating the palace of Versailles that we know today.[65]

Second, the king asked Le Nôtre to expand and refine the gardens and create an estate worthy of a royal capital. The devoted gardener set to work. Over the previous few years Le Nôtre had already transformed the Allée Royale from a simple grand avenue to a monumental one. At its origin in front of the château he replaced the elaborate embroidered parterre with an elegant flower-shaped pool. The "water parterre," as it was known, reflected Le Vau's new façade in its waters and alleviated the stone structure's heaviness in a way the original pattern could not. Farther down the Allée Royale he made the existing pool, which had hosted some of the most memorable events of the celebrations of 1664, into the famous Bassin de Latone, and beyond it, at the end of the original axis, he transformed the old Fountain of the Swans into the magnificent Fountain of Apollo. He now elaborated the garden sections north of the château into two elegant bosquets with flowing water and a Dragon Fountain and a larger Neptune Fountain. To the south of the château he established the Orangerie, with monumental staircases leading to it on each side.

The Hidden Pattern

Yet while all of these improvements were charming, tastefully done, and pleasing to the king, all of them were made within the bounds of the Petit Parc and did not change the overall feel and look of the gardens. Only one of Le Nôtre's projects at this time did that: the Grand Canal, begun in 1668 and completed in a remarkable three years. For the Grand Canal lay in its entirety outside the old boundaries of the garden, in the area that came to be known as the Grand Parc. Slightly less than a mile long and more than two hundred feet wide, the main arm of the canal is a direct continuation of the Allée Royale beyond its old terminus of the Fountain of Apollo. Just short of its midpoint it is intersected at right angles by a transverse canal of the same width, running 650 yards to the north and 450 yards to the south, creating the outline of a watery cross in the valley. With the canal in place, Le Nôtre more than doubled the overall size of the gardens and entirely changed their complexion. When viewed from the palace, the gardens no longer seemed like an enclosed space, but an open vista stretching out to the horizon.

We have already noted how the central axis at Vaux had turned that garden into a perspectival painting, with the statue of Hercules serving as its vanishing point. The Grand Canal now did the same for Versailles, but on a far grander scale. Someone viewing the garden from the middle of the palace (later the Hall of Mirrors) would look down the Allée Royale, past the great fountains of Latona and Apollo, on to the placid waters of the Grand Canal and beyond, to a point on the horizon marked by a pair of poplar trees. So vast was the estate, and so far were the trees from the palace, that on many days they could not be seen at all. And yet, for the "painting" that is Versailles, they were the vanishing point, where all the parallel lines—main axis, parallel axes, garden boundaries, etc.—met.

With the Grand Canal, Le Nôtre did much more than enlarge the size of the garden: he changed its very nature, turning it into a physical, three-dimensional geometrical space. The Petit Parc was, to be sure, geometrical, but it was so because of the careful arrangement of its parterres, bosquets, and fountains. The area surrounding the Grand Canal, in contrast, was far less intensively cultivated and for the most part took the shape of an open wood dissected by straight paths. From the château it looked like nothing so much as a natural forest, no different from others in the Île-de-France. Yet it did not matter: the Grand Parc was geometrical not because of human artifice, but by virtue of the universal laws of perspective. Like the inner space of a Renaissance painting crafted according to the principles of linear perspective, every point in the gardens, whether in a meticulously crafted parterre or a free-form wood, could be determined in relation to the observer and precisely calculated. The Grand Parc of Versailles, stretching to the horizon, may have seemed like rough, disorderly terrain, but this was a mere illusion: underlying the chaotic exterior of the woods was a precise and unyielding geometrical pattern.

Indeed, the idea that a rigid geometrical order pervades all of nature, even when it appears wild and chaotic, was written into Versailles's Grand Parc. The woods, dark and impenetrable though they appear from the château, are in fact crisscrossed by arrow-straight pathways that intersect at precise angles, creating a complex geometrical pattern that covers the entire valley. The pattern is nearly symmetrical, replicating itself in mirror image on both sides of the Grand Canal, and is composed of precise triangles: some of them equilateral, some of them right-angled, many of them isosceles triangles. Remarkably, quite a few of the paths that form this complex don't seem to lead from anywhere in particular to anywhere in particular. Their entire purpose seems to be to form the pattern itself.

The message was clear. To an outside observer the natural world appears impenetrable, just as the Grand Parc appears from the château.

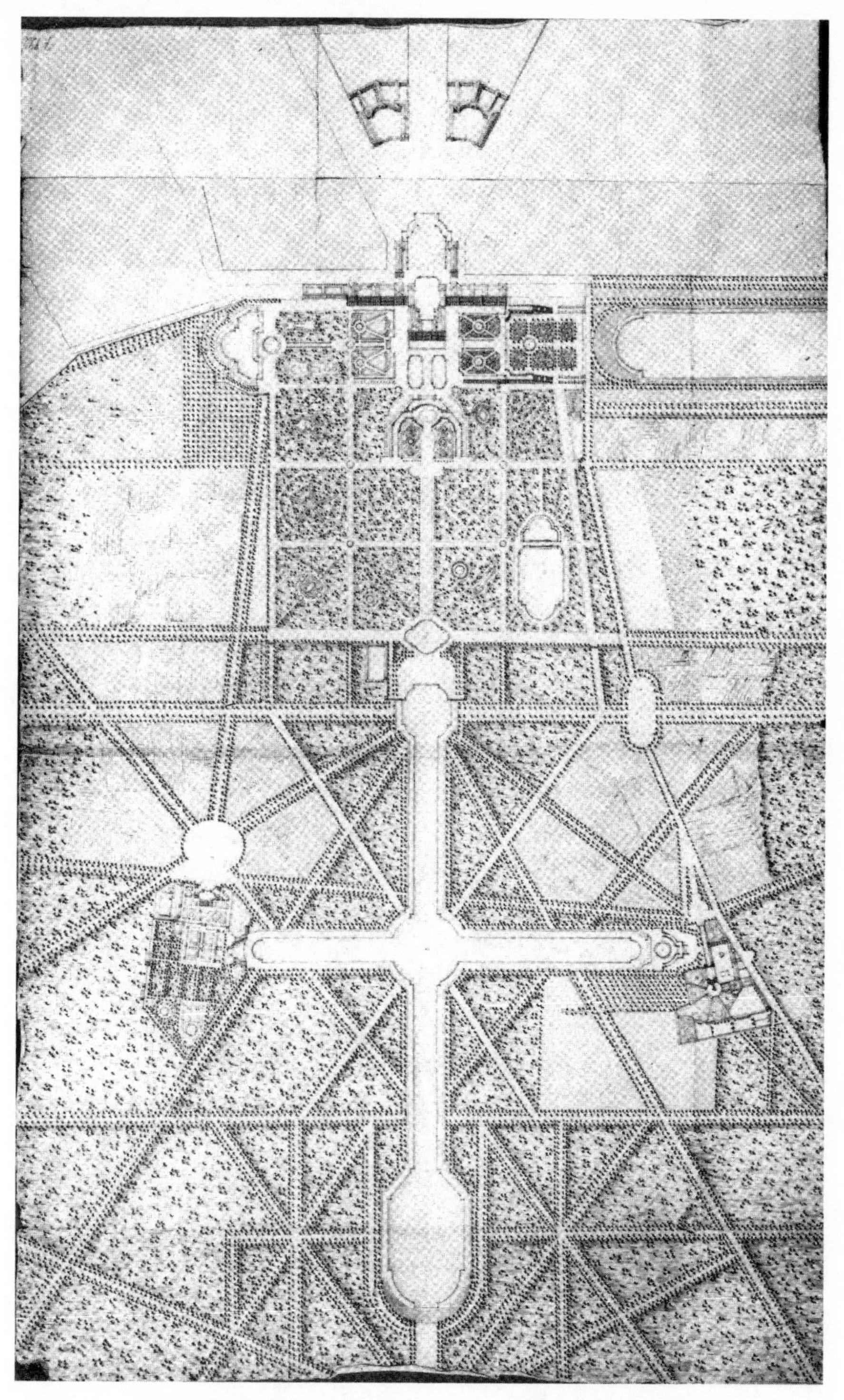

FIGURE 18: Plan of Versailles by Pierre Desgots, ca. 1669

For one immersed in nature and experiencing it at first hand it may yet appear overwhelming and mysterious, as the wooded valley may seem to one wandering its paths without a map or guide. Precisely the same is true of the human world, a tangle of overlapping groupings, interests, and power struggles that appears to defy pattern or system. And yet at Versailles one learns that the impenetrable chaos of the world, both natural and human, is nothing but an external, superficial guise. Just like Versailles's Grand Parc, where open woods overlay a precise geometric pattern, our rough, irregular, and diverse reality is in fact precisely and geometrically ordered. For the world, as Brunelleschi, Alberti, and Masaccio had taught two centuries before, is geometrical to the core.

It was a lesson that Louis XIV knew only too well. In 1669 he invited the Italian astronomer Giovanni Domenico ("Jean-Dominique" in French) Cassini (1625–1712) to become the director of the newly built Paris Observatory. Along with the normal taking of astronomical measurements that the post required, the king also set him an additional task: to accurately map and measure the territory of France.[66] Backed by the king's authority, Cassini set to work.

By the time it was completed in the 1790s the Cassini map of France had consumed the efforts of four generations of Cassinis, all of them directors of the Paris Observatory. They had produced what was then by far the most accurate map of any territory on Earth. But even the early effort of Jean-Dominique (also known as Cassini I) in the 1670s and '80s was far better than anything else available. When he presented his map to Louis in 1693, the king was distressed to discover that the Atlantic coast of France was farther to the east than previously thought, and that much of what was marked as part of his territory in older maps was in fact open water. "You have lost me a third of my kingdom," Louis protested to Cassini, but he resignedly accepted the map's verdict. For the Cassinis, from Jean-Dominique onward, had relied on the most accurate cartographic method available before the advent of aerial photography. It was known as triangulation.[67]

In principle, triangulation is a simple exercise in Euclidean geometry: once the distance between two points is known, the precise dis-

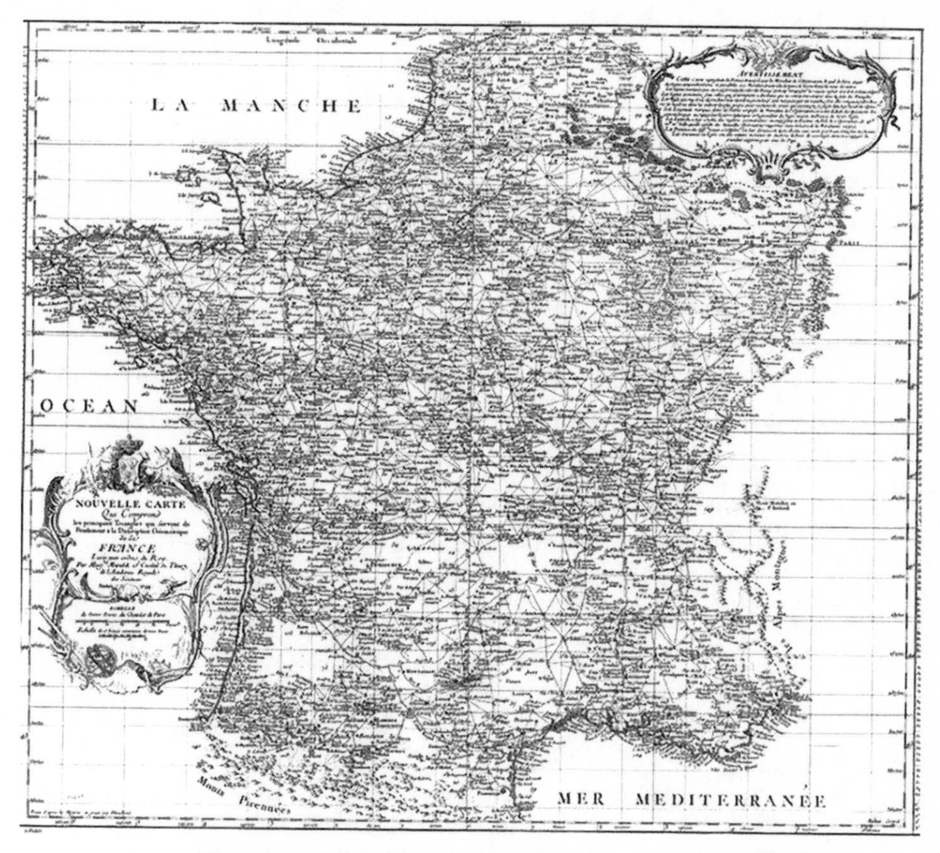

FIGURE 19: The triangulation of France. Map of France showing the principal triangles used in its creation.

tance to a third point can be calculated by measuring the angle at which it is viewed from each of the original points. The three points now form a triangle whose sides are all known, and each side can now serve as a baseline for calculating the location of a new point, thereby forming a new triangle. In this way, by constructing triangle upon triangle, one can calculate the precise distances and relative angles between any two measured points and gradually construct an accurate map of any territory. In practice the method is far more challenging: one needs to select prominent points that are easily distinguishable from others nearby to make sure that each point is clearly visible from the others, that it is physically—and sometimes legally—accessible, and that the heavy measuring instruments can be hauled up to each point without suffering damage. Finally, one needs to carry out the measurements at each point with the highest degree of precision. Even small errors in measurement, when compounded many times over, can make the resulting map effectively useless. The challenges were formidable, but no one faced them more doggedly than the Cassinis, who made the new map of France a family business for more than a century. In the process, they covered the entire territory of France with precisely calculated triangles.

As visitors to Versailles could not fail to note, Cassini's triangles had their counterparts in the Sun King's garden. Like the territory of France, the Grand Parc was systematically carved into regular triangular patterns laid layer upon layer until, taken together, they covered the entire garden. The patterns of the Grand Parc told of a deep geometrical order that underlay the seeming chaos of the world. The triangles of the Cassinis showed how this order could be revealed in the land itself, and put into the service of a rational central bureaucracy.

The Royal Eye

There are many messages encoded in the gardens of Versailles, lessons to be taught to a visitor at the park. Some of them have to do with the king's person. The Fountain of Latona, for example, tells the story of Latona (or Leto), who, having been seduced by Jupiter and given birth to his twins, Apollo and Diane, was fleeing the wrath of his companion, Juno. Stopping with her children to refresh themselves by a clear pool, she was set upon by uncouth herders who stirred the mud and made the water undrinkable. As punishment, Leto transformed the offenders into frogs—the moment captured in the fountain. The analogy to Louis's own childhood experience during the Fronde, when he and his mother were surrounded and insulted by rude commoners, only to emerge victorious in the end, could not be missed.

The didactic identification of Louis with Apollo was made even more explicit in the next great pool down the path of the Allée Royale, the Fountain of Apollo. Here the lesson is not so much about Louis the man as about Louis the king, the absolute ruler of France. The statue at the center of the fountain depicts Apollo in his chariot rising from the waves at dawn, accompanied by leaping dolphins and tritons blowing their conches. The Sun God is facing the château with the clear intention of soaring through the skies above the Allée Royale. His final destination in the early years was the Grotto of Thetis, situated just north of the old château, though ultimately demolished to make room for the great north wing of the palace. Incongruously, Apollo's daily journey at Versailles, from west to east, was the reverse of the natural course of affairs, but the message was nevertheless inescapable: the Sun King is the Sun God, illuminating his kingdom day by day as brightly as Apollo lights up the world.

There are other lessons as well: a fleet of ships sat at anchor in the Grand Canal, including, as John Locke reported, "a man of war of 30 guns, 2 yatchs and several other lesse vessels." These reminded visitors that Versailles stood for France itself, a great kingdom whose might spanned the oceans.[68] The menagerie, situated at the end of the southern arm of the Grand Canal, housed both domestic and wild animals such as lions, stags, an elephant, and a rhinoceros, as well as an aviary of exotic birds. It completed the picture of the gardens as a representation of the natural world, in which not only trees, shrubs, and flowers find their place, but animals as well.[69]

The Slave Fountain in a bosquet called l'Encelade depicts a giant fated to toil in the depths of the Earth, struggling to break his bonds. The giant is immensely strong, his rage palpable, his cry of defiance visible as a massive water jet shooting toward the heavens. And yet for all his might and fierce determination, his struggle is doomed from the start, for no one, not even the powerful, fierce, and oppressed, can challenge their place in the great hierarchy of being.[70] Cascades, in which water pumped up a slope rushes irresistibly downhill until it reaches a peaceful equilibrium, tell of how everything in the world strives to reach its own natural level. Terraces and monumental staircases leading from the valley up to the palace evoke the rigid, immutable hierarchy of the world.

All of these offered valuable lessons for a visitor to Versailles, and all of them were intended to glorify the king and enhance his rule. Yet no lesson was more deeply ingrained in the structure of the gardens, or more frequently repeated, than the underlying geometrical order of the rich wilderness of creation. It was evident in the elegant topiary that turned the trees along the pathways and parterres of Versailles into precise globes, cones, and obelisks. It was also the lesson of the popular bosquet dubbed the Labyrinth, which was specifically designed so that visitors would lose their way among its paths, mimicking the experience of confronting nature without map or guide. Yet hidden just below the chaotic surface was a clear, coherent, and orderly pattern: when walking through the Labyrinth one experienced wonder and confusion, but seen from above, or on a map, the unconquerable maze dissolved into a simple geometrical arrangement of rectangles, diagonals, and semicircles, all interconnected by straight lines. The confusion of nature, the Labyrinth told, was only surface-deep. Just beneath it lay the deep and unalterable order of geometry.

The same message was, in fact, implicit in practically all the bosquets that took up most of the area of the Petit Parc. Each bosquet—literally, "little forest"—occupied one of the square or rectangular plots that were created by the intersections of the east–west with the north–south avenues. From the outside they were practically indistinguishable, each presenting a thick and impenetrable wall of trees. But when one ventured inside, surprises awaited. One bosquest presented a graduated trio of fountains with water cascading from one to the next and jets shooting into the air; another presented a pool with a round island of yellow sand in the middle and straight paths leading outward like the rays of the sun; yet another, a golden triumphal arch and a fountain statue of France triumphant; and so on.

The effect on a visitor was—and still is—one of astonishment that such wonders are hidden within the monotonous and seemingly impenetrable woods, where one never knows what the next turn along the path will reveal. And yet, a glance at the plan of each and every bosquet shows that their design is in fact strictly geometrical, and the paths—so mysterious and unpredictable as one walks through them—follow straight lines and right angles, forming perfect circles, squares,

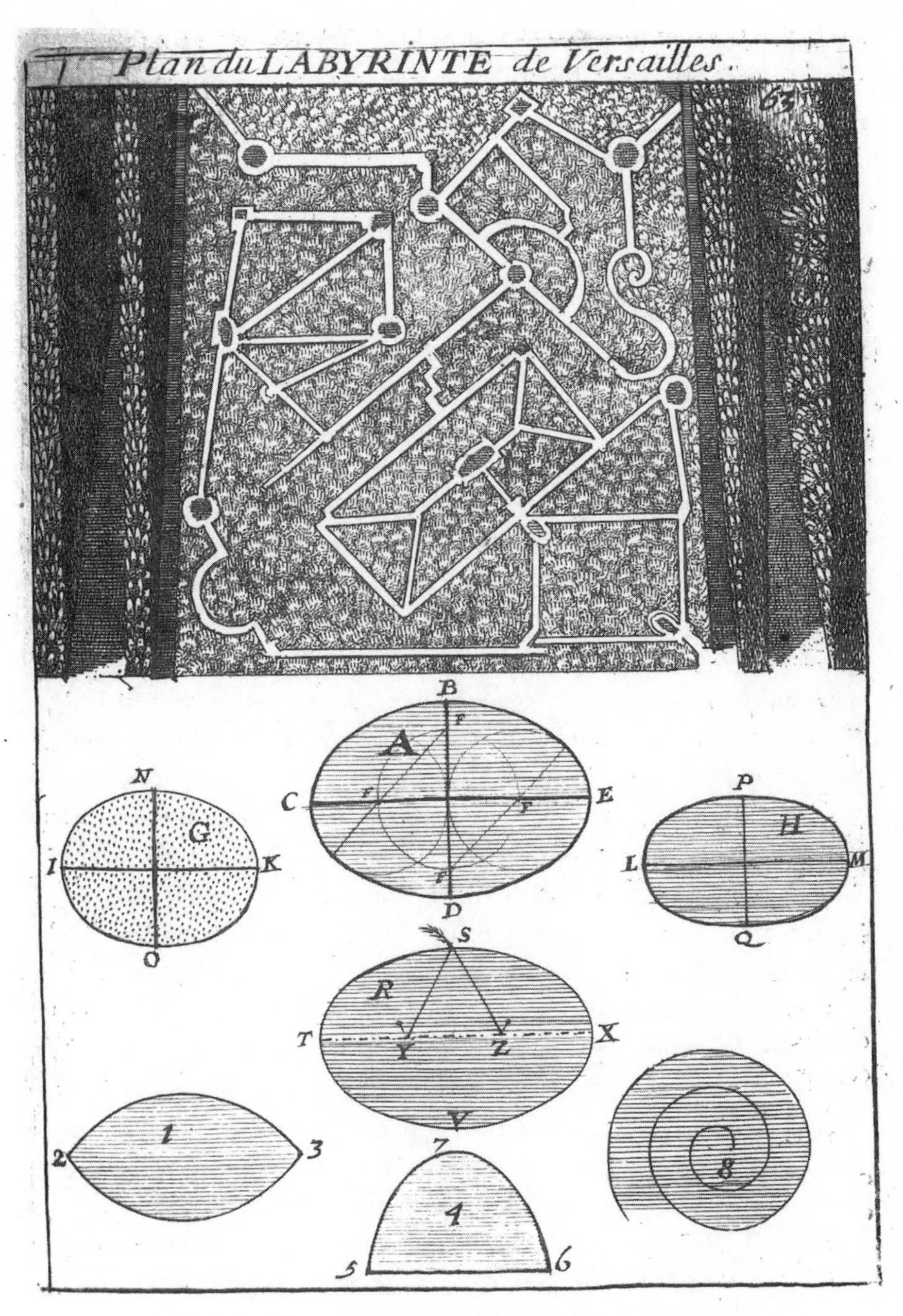

FIGURE 20: The geometry of the labyrinth bosquet at Versailles by Alain Manesson Mallet, 1702

and triangles. Just like the woods of the Grand Parc and the carefully groomed trees along the Allée Royale, the magical forests of the bosquets are grounded in universal geometry.

The message echoes through the gardens from one end to the other, but nowhere more so than in the geometrical paths of the giant Grand Parc. It is not just a geometrical order that the paths express, but a directed one, with a target clear in its sights. Four broad, arrow-straight avenues slice through the woods and converge symmetrically, two on each side of the canal, onto its eastern basin, the one closest to the Allée Royale. The two inner avenues are the longest in the garden, and together all four establish the overall pattern of the paths in the wooded Grand Parc. It is shaped like an arrow, with the canal itself forming the shaft and the converging avenues the pointed head. And it is aimed directly at the middle of the Petit Parc, upward through the Allée Royale, and on to the center of the palace and the king's bedchamber.

We have already seen how the crisscross pattern of the paths in the Grand Parc revealed a hidden geometrical order. Now, the overall arrow shape underscores a deeper truth about this order: it is hierarchical through and through. The elaborate pattern of triangles that pervades the park rises up, layer upon layer, to constitute the great arrow that points uphill. At its apex, towering high above the woods of the great park and held in place by a rising tide of geometrical patterns aimed at its heart, is the royal palace. The supremacy of the king and his rule, it is now clear, is the culmination of the deep geometrical structure of the world itself. Just as geometrical order is eternally hierarchical, so is the natural social order of France. And just as the truths of geometry are undeniable and irresistible, so is the rule of Louis XIV. At the Grand Parc one learns that the absolute monarchy of the Bourbons is no coincidence of history, no accident of power politics. It is the inescapable expression of the geometry of nature itself.

The universal principles of geometry, as presented at Versailles, were the deep foundations of royal supremacy. Like geometry, the truth, hierarchy, and order of the king's rule could not be denied, undermined, or overturned. This was clear to anyone raising their eyes from the gardens to look up at the château: the entire construction of the park pointed to the royal palace at the top of the hill, supporting and buttressing its exalted position. Meanwhile, the same geometrical structure also facilitated something else: the uninhibited flow of power downward from the palace to every corner of the park, and by extension to the kingdom.

Bishop Bossuet, Louis's official theorist of royal absolutism, compared the king's power in his kingdom to God's power over creation: "The power of God makes itself felt in an instant from one end of the world to the other; royal power acts similarly through all the realm. It keeps the whole realm in order, as God keeps the world."[71]

And that is indeed the power that revealed itself to the king as he looked down from the royal palace to the gardens that opened up below him. Straight, uninterrupted lines spread out from the palace to the farthest ends of the valley like the rays of the sun. In the geometrical landscape of the gardens of Versailles no corner was so remote or so obscure as to be hidden from the all-encompassing royal gaze or sheltered from the reach of royal power. It was a pattern that no doubt appealed to Bossuet, who had structured his most important work, the *Politique*, as a geometrical treatise. In such a land the will of the king, just like God's word, would indeed make itself felt everywhere in an instant.

In the kingdom of Versailles, power flowed in both directions. It flowed upward, from the gardens to the palace, demonstrating that royal supremacy was founded on universal and unchallengeable geometrical principles. And it flowed downward, from the palace to the gardens, ensuring that the king's sovereign will reached every corner of his kingdom without hindrance or opposition. It was the ideology of Louis XIV's absolutism written into the landscape. Royal supremacy is absolute, and royal power limitless. The reason was simple, and to anyone walking the garden paths of Versailles, self-evident: both were grounded in the absolute, unchallengeable, limitless, and irresistible science of geometry.

PART III

≡

The Sun King's Touch

6.

BEYOND VERSAILLES

Twilight of the Sun King

When it came to publicly presenting his own power and position, Louis XIV was not one to leave anything to chance. "There is nothing in this matter that is unimportant or inconsequential," he admonished the Dauphin in his *Mémoires*, and throughout his reign he never compromised on any point of ceremony or public display that might enhance the Sun King's stature. And so even at Versailles, where every path, stone, shrub, and tree seemed to proclaim the glory of the Sun King's reign, Louis was not content to let visitors wander the park and draw their own conclusions.

To make sure that all would experience the park as they should, the king authored a guide to the gardens that specified in precise detail where one should go and in what order, and what should be observed at every point. Such was the importance he assigned to the matter that between 1689 and 1705 the king authored no fewer than six versions of the guide, issued most likely in preparation for the visits of foreign potentates or diplomatic missions. Each version was only a slight refinement of the previous one, and all took the visitors along roughly the same route: a clockwise circuit through the bosquets of the Petit Parc, beginning and ending at the center of the palace, the point of origin of the Allée Royale. For the Grand Parc, Louis's guide offered separate excursions to the Grand Canal, the menagerie, and the intimate palace of the Trianon.[1]

It was a fitting tour for a geometrical garden. Nowhere in the guide does the king offer information about the plants of the garden, about the vast hydraulic system supporting it, or about the mythological themes of the ubiquitous statuary and fountains. Instead, the guide provides detailed instructions of where to turn at every juncture, where to stop to take in the view, which direction to look in at each stop, and what to observe. Far from an amusing sightseeing excursion, the king's tour of the park was a ceremonial dance in which every movement, stop, and turn of the head is carefully choreographed. For it was not the verdant beauty of the gardens, nor their mythological representation of the Sun King, nor even the lavish display of wealth and engineering prowess that mattered most to Louis XIV. It was, rather, their precise geometrical order, which governed everything within them. By carefully following each step in the king's elaborately structured tour, visitors would themselves become part of the garden's deep geometry. Willing or not, they were conscripted into the grand edifice of order and hierarchy designed in every detail to uphold the monarchy and proclaim the supremacy of the Sun King.

That the king succeeded in impressing his message of incomparable power and grandeur on his visitors there is no doubt. A long list of royals, dignitaries, and ambassadors who were given guided tours of Versailles attested to that. The Russian ambassador, who was led through the gardens in 1681, proclaimed that "there had never been [anyone] on the earth but Solomon and the king of France who had appeared with so much grandeur."[2] A member of a delegation from Algiers, which visited three years later, declared upon seeing the Grand Canal that "the sea of Versailles was the sea of the Emperor of the World."[3] And ambassadors from faraway Siam (Thailand), who in 1686 were treated to a multiday tour of the gardens, declared that "it was impossible to carry magnificence beyond that of the King."[4] The geometrical gardens had become, for Louis XIV, a powerful tool of diplomacy, establishing his ascendancy both at home and abroad.

Initially, events seemed only to confirm the Sun King's confidence in his glorious destiny. In those early decades of his reign Louis was without peer among European monarchs, the most dynamic and inspiring sovereign on the Continent. His armies went from triumph to triumph, and France rose quickly to become the leading power in Europe. For Louis these dazzling successes were a clear vindication of his faith in absolute royal power, and they encouraged him to assert his status and authority with ever greater vigor.

Progress at home seemed equally promising. After years of gradually transferring the functions of government from Paris, Louis officially moved the court to Versailles in 1682, making it his primary residence and the administrative center of the kingdom. This made Versailles home not only to the sizable court that moved with the king from the Louvre, but also to noblemen and noblewomen from across France who swarmed the town and palace in a rush to be near the king. Anyone aspiring to make their mark in the kingdom of Louis XIV was required to be present at Versailles. A century and a half before, Francis I had been constantly on the move, asserting his royal claims by traversing the country to visit the estates of his great nobles. At Versailles the tables were finally turned: now it was the great nobles who were traveling, leaving their country estates behind to join Louis in his glamorous seat.

Yet in the very years that the court of Versailles was at its most radiant, the years when visitors to the court were conducted through the gardens and imprinted with the glory of Louis XIV, storm clouds were gathering over the land of the Sun King. Much to Louis's dismay, the same policies that had seemed to produce an unbroken string of successes in his early years now seemed to produce only stalemate and failure.

The more he moved to assert his authority, the more he pressed his enemies and set out to crush them, the more they drew together to resist his designs.

At home the king set out to end the religious toleration of Protestants. In 1685 he revoked the Edict of Nantes, which had guaranteed the Huguenots freedom of worship since the days of Henry IV. To Louis, who detested all forms of special privilege that came between him and his subjects, it was an obvious decision and a critical step in establishing his absolute rule. Yet while solving one problem, the revocation only raised a host of new ones: the brutal suppression of the Protestant faith and the persecution of its adherents discredited the king and undermined his standing around Europe. It alienated all Protestants—including those who had previously been his allies—thereby making all the king's future ambitions, military as well as diplomatic, immensely more difficult. In France, meanwhile, many Huguenots continued to practice their religion in secret while dissembling Catholic observance, creating insoluble problems for the Church. Many others were driven into exile, infusing the countries in which they settled with a bitterness toward the king that would haunt him for the rest of his days.

Abroad the king fared no better. The revocation of the Edict of Nantes isolated Louis XIV's France diplomatically, but it was Louis's own unbridled aggressiveness that united his enemies into a powerful military alliance. In the 1680s he ordered the naval bombardment of Genoa and the land bombardment of Luxembourg, and without provocation he sent his armies to lay waste to the western German principality of the Palatinate. But when he tried to intervene in favor of the recently deposed James II of England, he provoked a war against a grand alliance of England, Holland, Prussia, and Austria that would last nine years and cost Louis all of his conquests of the previous decade. Louis was unchastened. Only a few years later he tried to take advantage of a dispute over the succession to the crown of Spain to occupy the Spanish Netherlands and open the way for a union of the French and Spanish crowns. In the bloody war that followed, France once again found itself isolated against all the other major powers of Europe, and this time Louis's armies fared even worse. Crushing defeats in the battles of Blenheim (1704), Ramillies (1706), and Oudenarde (1708) and an inconclusive bloodbath at Malplaquet (1709) brought France to the point of capitulation. Only dissension among the allies and a last-minute patriotic surge by the French in the face of foreign invasion saved the Sun King from utter humiliation.

When Louis XIV died in 1715, the glorious early years of his reign were but a distant memory. Decades of warfare and defeat had left the state finances in ruins, the peasantry overtaxed and impoverished, villages and towns depopulated, and large swaths of territory ravaged by foreign armies. The king whose assumption of personal rule in 1661 was greeted by a wave of national euphoria died hated by his exhausted people, who held him responsible for a seemingly unending string of calamities.

The Bourbons would continue to rule France for another seventy-four years, but they never recovered from the spectacular failure of Louis XIV's ambitions. Never again would French kings rise so high in the esteem of their countrymen and of foreigners, never again would the royal court shine so brightly, and never again would it be the guiding light of the cultural and intellectual life of France and of Europe. The eclipse of the Sun King was not his personal failure alone: the vision he embodied of a harmonious kingdom, flourishing under the rule of a monarch whose absolute powers were grounded in the rational order of the world, had failed with him. To Frenchmen, the promise of Versailles that absolute royal rule would lead to peace and prosperity at home and power and glory abroad was revealed as a monstrous sham.

Versailles Beyond Versailles

Outside of France, however, views of Versailles were decidedly different. For even as Louis XIV's fortunes faltered in his final decades, the geometrical gardens of Versailles never ceased spinning their magic over friend and foe alike. Forever identified with the Sun King, and inseparable from his rule, they became the object of envy and imitation by those who dreamed of replicating his glory. It was not long before every great prince and princess in Europe was racing to create their own gardens of Versailles in their capital. So closely were Louis's gardens associated with royal supremacy and glory that no king or emperor seeking status and recognition on the Continent could afford to do without them.

First in line to attempt to create their own Versailles were none other than the kings of England. This might seem surprising in view of the bitter enmity between the two kingdoms in the latter part of Louis's reign, and the subsequent development of the naturalistic English gardening style as a conscious rebuke to Le Nôtre's geometries. But in 1662 Charles II, newly restored to the English throne, relied heavily on Louis's diplomatic and financial support to maintain his crown. Of Versailles he most likely heard from his sister Henriette, who was Louis's sister-in-law and intimate friend. Seeking to bask in the Sun King's reflected glory, Charles asked his French "cousin" for Le Nôtre's services in creating his own gardens at Greenwich Park. Louis, perhaps reluctantly, agreed, but in the end the very busy gardener never made the journey to England. Instead he produced designs for the park, which were then implemented by Charles's landscapers to the best of their abilities.[5] Thirty-six years later, during a lull in the bloody wars, it was William of Orange, King of England, who requested the famous gardener's help in creating a worthy park at Windsor Castle. The octogenarian Le Nôtre was too old to travel, but he once again provided some general designs and sent his nephew, Claude Desgots, to supervise the work.

The grounds of Greenwich and Windsor, as seen today, are both charming royal parks that show residues of Le Nôtre's touch. In both a broad, arrow-straight central avenue runs from the royal palace to the outer reaches of the park, while other straight allées converge on it from both sides, creating elegant geometrical patterns. But neither park would remind one even remotely of the brilliance of Versailles. This is probably appropriate, for the kings of England could only dream of the royal magnificence attained by Louis XIV. Following the English Civil War and the royal restoration of 1660, English kings had to contend with a powerful Parliament that kept a tight grip on the purse strings and was deeply suspicious of the power of kings. Gardens on the scale of Versailles, with their message of absolute royal supremacy, would never be allowed in England.

But elsewhere things were different. On the eastern borderlands of Europe, Tsar Peter the Great (1672–1725) was eager to establish Russia as a great power and himself as equal to the greatest princes on the continent. An autocrat unburdened by elected assemblies, Peter built himself the beautiful new capital of St. Petersburg on the shores of the Baltic Sea. Along the coast, some twenty miles west of the city, he created a pleasure retreat and invited one of Le Nôtre's disciples, Jean-Baptiste Le Blond (1679–1719), to design the palace and gardens and supervise their construction. He named it, predictably, "Peterhof" (Peter's Court). Le Blond set to work with energy and flair, and by the time of the Tsar's death in 1725, the complex was largely completed. Perched on top of a bluff, the royal palace overlooked a vast expanse of parterres, bosquets, and woods, all geometrically arranged along straight boulevards. It was indeed, as Peter had intended, a "Russian Versailles."

The "Spanish Versailles" was built around the same time as the Peterhof, at La Granja de San Ildefonso, near the town of Segovia, north of Madrid. It was the creation of King Philip V (1683–1746), who, having miraculously survived the War of the Spanish Succession, decided to build himself a palace that would erase the bitter memories of the struggle. Created with the aid of French architects and landscapers, the gardens of La Granja were built on a truly royal scale, complete with a canal, parterres, bosquets, and a labyrinth, all arranged in distinct geometrical patterns. It was modeled self-consciously on Versailles, and like the Sun King's palace it soon became the main seat of the court and the center of the royal administration.

Philip was Louis XIV's grandson, and it is therefore not entirely unexpected that he would choose to create his own Versailles. But Louis's most bitter enemy, the Habsburg Holy Roman Emperor Leopold I (1640–1705), did much the same. Leopold fought three brutal wars against Louis XIV, but far from souring him on all things French it only inflamed his envy of the Sun King's magnificence. In 1695, in the midst of the bloody struggle that pitted the emperor and his allies against

France, Leopold commissioned Jean Trehet (1654–1740), another of Le Nôtre's disciples, to design the gardens for the imperial palace of Schönbrunn on the outskirts of Vienna. The old estate had been severely damaged during the Ottoman siege of Vienna some years before, and the emperor was intent on rebuilding it in a style worthy of his exalted station. And that, to him, meant the style of Versailles.

Trehet did not disappoint. He created a broad and straight central avenue leading from the palace, surrounded on each side by symmetrical parterres with elaborate patterns. Flanking the central axis on each side were elegant bosquets traversed by straight alleys that formed regular geometrical patterns and converged from all directions on starlike plazas. A second burst of construction under Empress Maria Theresa (reigned 1740–1780) expanded the gardens farther south, adding large basins, fountains, and a triumphal *gloriette* on a hill opposite the palace, but did not interfere with the gardens' overall design. They remained classical geometrical gardens, patterned after Versailles but glorifying the Bourbons' rivals, the Habsburg emperors.

And this was indeed the most direct legacy of Versailles. From Spain and England to Austria and Russia, every great prince rushed to create his own replica of Le Nôtre's gardens. If the original Versailles presented Louis XIV's rule as an integral part of the universal order, then clearly his rivals could not let things rest at that. Each of them now perceived an urgent need to assert himself (or, in Maria Theresa's case, herself) as just as powerful, irresistible, and inevitable as the Sun King, and Versailles demonstrated how it was done. The kings of France were the first among the great princes of Europe to make geometry the ideological foundation of their power, and none did so on a more magnificent scale or to greater effect than Louis XIV. But by the eighteenth century other ambitious European monarchs had caught on, creating their own geometrical kingdoms around their own palaces. There is irony here: the gardens of Versailles, created to present and glorify the rule of the one and only Sun King, ended up replicating themselves from one end of the continent to the other, in the service of rival monarchs. The language of geometry, it turned out, could not be contained.

While the cloning of Versailles by kings and emperors throughout

Europe was the most obvious legacy of Le Nôtre's gardens, it was also the most short-lived. In 1830, following four decades of revolution and war, the last of the Bourbon kings of France was chased from his throne, and by the early twentieth century most of the great European monarchs had followed in his wake. Some came to a brutal and bloody end, others abdicated under pressure, while a few were fortunate enough to survive—at the cost of surrendering all political power. The great palaces and gardens built to legitimize mighty autocratic monarchies became museums frozen in time, administered by popular regimes that would make their creators shudder. Louis XIV, Peter the Great, and Leopold I would no doubt be turning in their graves had they seen the sunburned tourists roaming daily through their magnificent royal estates, now turned into diversions for the masses. And so, while some of the grander geometrical gardens do survive—at Versailles, in Vienna, in St. Petersburg, at La Granja—their power to order social and political hierarchies has been neutralized.

The Rational State

If the legacy of the gardens of Versailles had been limited to their replication in European capitals, then we would have to conclude that their impact had come to an end a century ago, with the fall of the last great European monarchies. But the power of Versailles extended beyond its explicit support for absolute monarchies. More broadly, the geometries of Versailles stood for a vision of the modern state as orderly, rational, scientific, and bureaucratic, a vision that is as powerful today as it was in the age of Louis XIV.

The strict orderliness of Versailles, and the broad, straight boulevards reaching to every corner of the park, lent support, as we have seen, to Louis's absolutist rule. But they also represented and forcefully buttressed the centralized bureaucratic state that Louis's minister Colbert was working tirelessly to create. To the Sun King's contemporaries, the unobstructed lines emanating from the palace were a clear representation of his irresistible power and unchallengeable rule. But they were also representations of an all-powerful central administration reaching out to every corner of the land, sending out emissaries and instructions, and receiving in return a steady stream of revenue and information.

It is this vision of a powerful state, rationally administered by an efficient central bureaucracy, that made even as fierce a critic of royal absolutism as Voltaire (1694–1778) into an admirer of the Sun King. In Louis XIV's France, he wrote, "human reason in general was brought to perfection," and "a general revolution took place in our art, minds, and customs, as in our government."[6] The king "had only to command and administrative conquests followed as swiftly as his conquests in the field." Voltaire then went on to praise the king's administrative triumphs in constructing defenses, rebuilding seaports and the navy, reforming the army, reviving overseas commerce, and constructing huge buildings, which "gave work to thousands of men, fostering all the arts that architecture brings with it."[7] There was no limit, it seemed to Voltaire, to what a powerful central authority could accomplish through an efficient bureaucracy.

And therein lies the true legacy of the gardens of Versailles. The rigid social hierarchies represented in the gardens have lost their allure and have been replaced by the ideal (if not always the practice) of egalitarianism and social mobility. The absolutism of hereditary monarchies, glorified at Versailles, has ceded its place to the rule of democratic assemblies elected through universal suffrage. But Versailles's vision of the modern state as one governed by a powerful, rational, and efficient central bureaucracy survived, and it is still very much with us. And wherever this vision is publicly presented and reiterated by modern governments, echoes of Versailles linger still.

Little wonder then that the geometrical ideal of an efficient, rational state found expression not only in the gardens of royal palaces, but also on the bustling streets of capital cities—the homes of state bureaucracies. The first European city to be recast in geometrical form was likely Rome, where Renaissance popes created a central administration that would serve as a model for the national monarchies. Already in the 1530s architects working under the patronage of Pope Paul III experimented with some of the design elements that Le Nôtre would perfect more than a century later. They not only favored grand linear boulevards, but also designed and built the first examples of the "trivium"—the pattern of three straight and broad avenues converging on a central point. The dramatic scheme, later made famous by the approach to the palace at Versailles, can still be seen in Rome's Piazza del Popolo, where the central Via del Corso is flanked by arrow-straight boulevards known today as the Via del Babuino and Via di Ripetta.[8] Half a century later Sixtus V (pope from 1585 to 1590) commissioned the architect Domenico Fontana (1543–1607) to expand the isolated geometrical motifs into a grand urban plan for the Eternal City. True to his instructions, Fontana produced a design dominated by long, broad, and straight roads ranging from one end of the city to the other, and forming elegant plazas at their intersections. Yet the overall plan lacks the symmetry and balance, as well as the ideological coherence, that French royal gardeners would develop and refine in the following century. Those, along with some key design innovations, such as star-shaped plazas, would be perfected in the gardens of Vaux and Versailles.[9]

If Renaissance Rome prefigured some of the geometrical elements made famous decades later by Le Nôtre, it was London that was the first city to feel the direct impact of the great gardener's work at Versailles. This came about because in 1666, only a few years after Charles II invited Le Nôtre to design the royal park at Greenwich, London was struck by the most destructive fire in its long history. For four days in early September the fire raged, scorching an estimated 13,500 structures, including St. Paul's Cathedral, Henry VIII's Bridewell Palace, the Royal Exchange, the Custom House, and eighty-seven parish churches. By the time the fire abated, most of old medieval London, from the Tower in the east to Temple Bar in the west, had been transformed from a crowded, bustling city into a blackened wasteland of ash and debris.

To the city's residents the Great Fire of London was an unmitigated disaster. While according to official figures only six people died in the conflagration, the true death toll was almost certainly far higher, and the commercial and economic damage was incalculable. But to the king the destruction also presented a dramatic and unexpected opportunity. Only six years after his restoration by Parliament, and still struggling to establish his legitimacy and authority, Charles saw a chance to transform

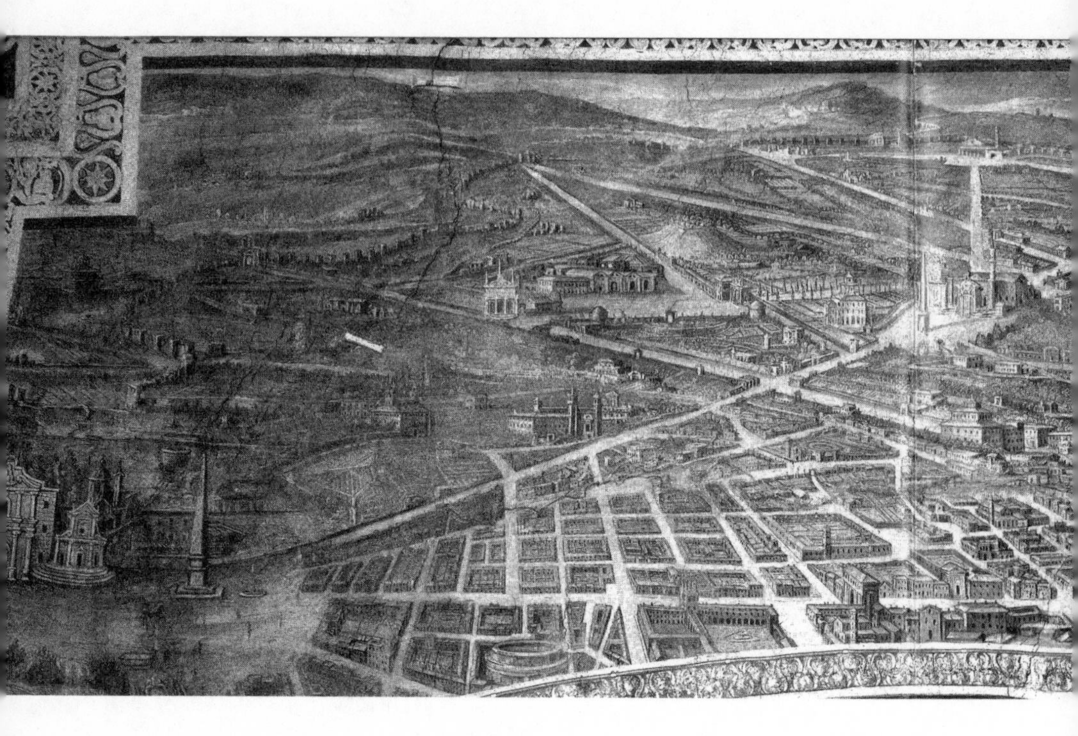

his fledgling rule. If London, a busy and crowded but admittedly rather dreary city, could be rebuilt in a style worthy of a great king, then Charles would no longer be the poor relative of European monarchs, subject to the whims of an elected assembly. A brilliant capital would go a long way toward legitimizing his royal claims in the eyes of his subjects at home, and his crowned rivals abroad.

And so it was not long after the last embers were extinguished along the Thames that the king began entertaining proposals for the rebuilding of London. While the total number of designs submitted to royal inspection is not known, eight survive, whose authors included some of the leading men of science and letters in the realm. Most proposals, including those by the cartographer Richard Newcourt (died 1679) and Robert Hooke (1635–1703), professor of geometry at Gresham College and curator of experiments for the Royal Society, took a practical approach, proposing variations of a rectilinear grid pattern that was friendly to commercial and residential buildings. But the two most dramatic

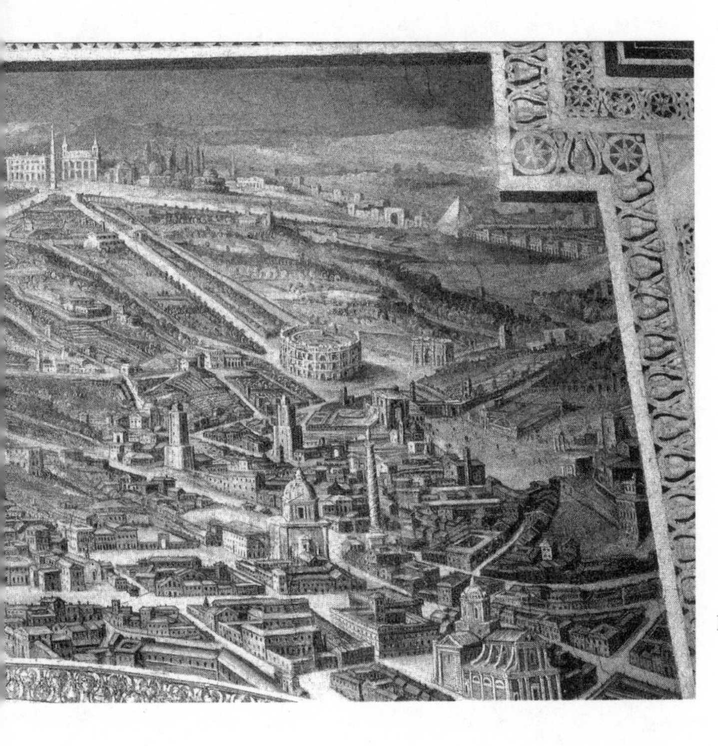

FIGURE 21: Fresco in the Vatican Library of Domenico Fontana's plan for Rome, ca. 1590

proposals took their inspiration directly from Le Nôtre, re-creating the elegant and hierarchical geometries of Versailles on the streets of London.[10]

That John Evelyn (1620–1706), a founder of the Royal Society and a famous diarist to later generations, would find inspiration for his urban design in gardening is hardly surprising. Evelyn spent almost fifty years writing and rewriting (though ultimately never publishing) a masterwork on the art of gardening that he called *Elysium Britannicum, or The Royal Gardens.*[11] A royalist during the Civil War, Evelyn was a believer in monarchy and an admirer of the geometrical style of gardening practiced in the palaces of the kings of France. Indeed, in one passage of the *Elysium* he suggests that the maxim Plato allegedly inscribed above the entrance to his Academy should with equal justice be set over his garden: "Let no one ignorant of geometry enter here."[12]

Evelyn's plan for London was, accordingly, the very image of a royal geometrical garden. At the center of the design is a grand oval plaza, in

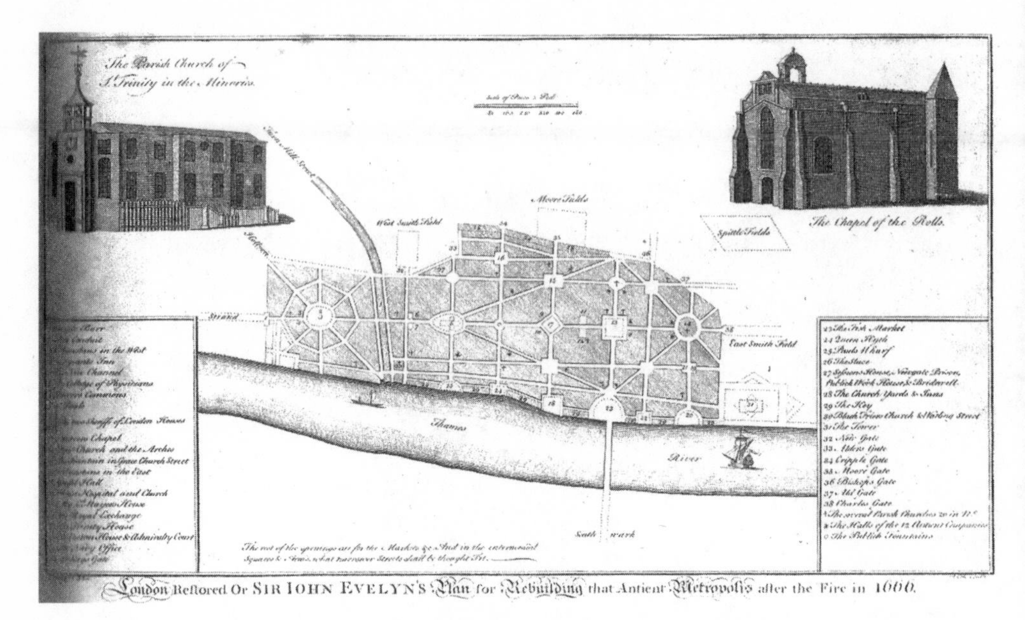

FIGURE 22: John Evelyn's plan for London, 1666

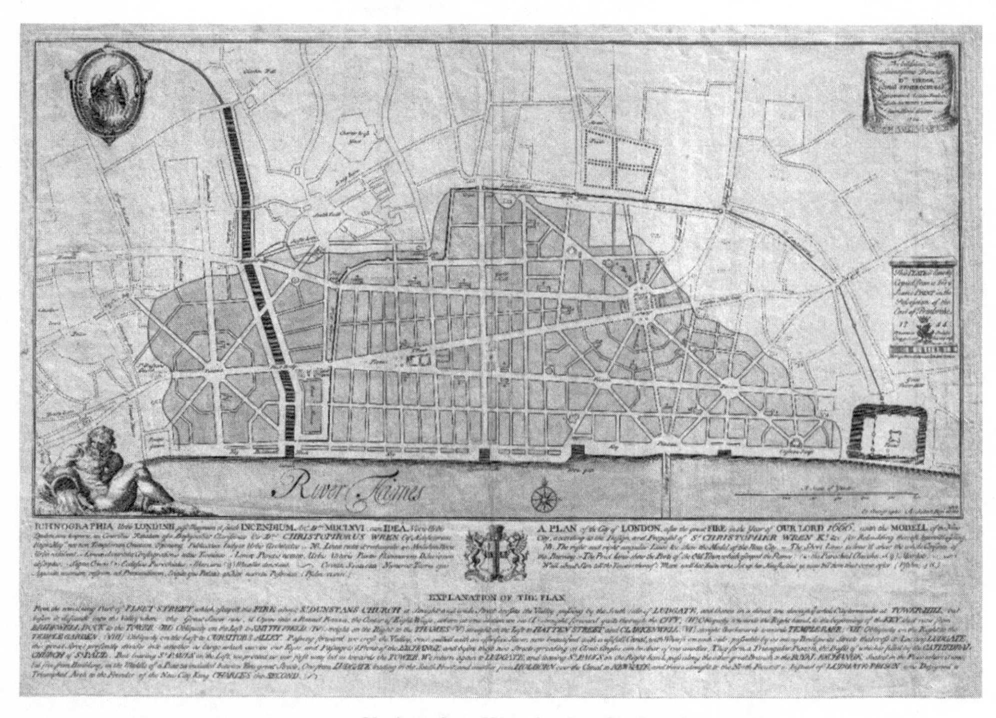

FIGURE 23: Christopher Wren's plan for London, 1666

which St. Paul's Cathedral would be rebuilt. Three broad boulevards converge on the plaza from the east, forming a trivium and connecting the cathedral to the other civic and ecclesiastical hubs: the Lord Mayor's house, the Custom House, the main churches, the fish market. The central axis then continues westward to a large octagonal plaza, with streets radiating from it, forming the shape of an eight-pointed star. Meanwhile streets running north to south intersect with the main east–west axis at roughly regular intervals, forming local squares and plazas at each intersection. It was almost as if Evelyn was examining a map of Versailles, and in particular Le Nôtre's plan for the Grand Parc, and decided to imprint it on the streets of London.

Christopher Wren's (1632–1723) design for London diverged a bit more from the Sun King's garden. Like Evelyn he made St. Paul's Cathedral a centerpiece of his plan, and the fact that he too proposed a grand octagonal star-shaped plaza at the western edge of the design area makes it hard to believe that the two did not, to some extent, collaborate. But Wren establishes an additional key node at the Royal Exchange, northeast of the cathedral, with streets radiating from it in no fewer than ten different directions. This creates a somewhat unbalanced design, with three main and several subsidiary east–west roads, some at oblique angles to one another. The regularly spaced north–south streets are replaced, on the eastern side of the planned area, by diagonals that intersect the main thoroughfares in large plazas. Whatever its functionality, Wren's design does not quite match the strict symmetry of Le Nôtre's gardens or Evelyn's city. Yet its grand axis, star-shaped plazas, and broad arrow-straight avenues converging on key centers of power all attest to the unmistakable imprint of Versailles. Wren's London, like Evelyn's, is a vision of a grand geometrical capital inspired by the Sun King's gardens.

Hooke's and Newcourt's proposals for the rebuilding of London were hardheaded and practical, suitable for a city that served as one of the great commercial hubs of Europe. Evelyn's and Wren's proposals were elegant, beautiful, and ideologically laden, proclaiming the power and grandeur of the kingdom and the king. But in the end, none of this mattered. Faced with a web of lawsuits and property claims over the devastated area, Charles was helpless. It is quite possible that Louis XIV, if faced with a similar challenge, would have sliced through the tangle by the sheer force of his personality and his absolute royal powers. But Charles, despite his grand hopes, was not Louis, and he soon had to concede defeat. All the plans, the practical as well as the grand, were set aside, and the city rebuilt as closely as possible to its original layout, each new street paved over the ashes of the old one. The only designer to benefit from the collapse of the king's ambition turned out to be Christopher Wren: over the next several decades he designed and built the new St. Paul's Cathedral, and no less than fifty parish churches in the fire-devastated area—all of them in an elegant, well-proportioned classical style.

While London's flirtation with the geometrical patterns of Versailles ultimately proved abortive, other capitals soon took the lead in integrating elements of the Sun King's garden into their urban landscape. This was no simple matter. The capitals of Europe were, for the most part, densely built and crowded cities whose street patterns, especially near their centers, had been set for centuries. This meant that any attempt at reconstruction had to contend with the existing urban patterns while taking advantage of whatever opportunities presented themselves. Things were simpler, however, when a new capital was built from the ground up. And such, in fact, was the case with the shining new Russian capital built at the turn of the eighteenth century on the shores of the Baltic Sea.

Peter the Great had grand ambitions for his capital. Named—perhaps inevitably—St. Petersburg, the city would express the hierarchy and order that governed the Tsarist autocracy, just as Versailles spoke for the ingrained hierarchy of Louis XIV's absolutism. Yet, also like Versailles, it would present a vision of reason and progress under state auspices, which the Tsar hoped to bring to his vast but backward empire. Initially Peter called upon Le Blond, Le Nôtre's disciple who also designed the Peterhof gardens for him, to draw up a master plan for the new capital. A gardener to the core, Le Blond proposed creating an entire city that would be nothing but a geometrical garden. In Le Blond's vision, St. Petersburg would be composed of a network of straight, broad boulevards intersecting at precise angles, large open squares with monumental statuary, and elegant formal parks with geometrical outlines, all nestled within walls that formed a precise oval. Le Blond's plan was ultimately rejected as impractical, but even the city as eventually built bears the clear imprint of Versailles. In particular, the trivium of grand boulevards converging like an arrowhead on the Admiralty building near the Tsar's Winter Palace are a direct echo of the three avenues that run through the town of Versailles and the broad allées that traversed the Grand Parc.[13]

More than a thousand miles southwest of St. Petersburg, the kings of Prussia were also exerting their autocratic power to modernize a militarily potent but backward kingdom. The rulers of Brandenburg-Prussia had been known as "electors" since the Middle Ages for their role in picking the Holy Roman Emperors, and their capital, Berlin, was centuries old. But after 1701, when the elector Frederick III declared himself king of Prussia, he and his successors were determined to put their stamp on the capital's urban plan and make it worthy of their new royal station. First and foremost in their designs was the creation of a royal boulevard in the heart of the city.

Unter den Linden, the avenue that became the ceremonial center of royal and imperial Berlin, began life in the seventeenth century as a simple country road and parade ground for the electors' troops. But beginning in the eighteenth century and continuing through the nineteenth, the electors' successors—the kings of Prussia and then the emperors of Germany—built Unter den Linden into one of the grandest boulevards in all of Europe. To anyone passing through the avenue, then as now, the echoes of Versailles are unmistakable. It began at the king's *Stadtschloss* (city palace), just as Versailles's Allée Royale began at the center of the royal palace. And just as the Allée Royale led straight to the dazzling Fountain of Apollo at the other end of the Petit Parc, so Unter den Linden—broad, tree-lined, and arrow-straight—led directly to the triumphal arch of the Brandenburg Gate. Along the way it passed the city's arsenal and key cultural and state institutions that were built on either side over the years: the State Library, State Opera, national memorials, museums, and Humboldt University. The key

government ministries stood nearby, ranged along the Wilhelmstrasse, which intersected with Unter den Linden at a right angle just before it reached its apex at the Brandenburg Gate. By the nineteenth century Unter den Linden had become a symbol and embodiment of the power and grandeur of the kings of Prussia and of the efficient bureaucratic state they had built.

A different approach, involving an original variation on the linear

FIGURE 24: **An aerial view of Berlin's Unter den Linden leading to the Brandenburg Gate (1931)**

patterns of Versailles, was tried by the Habsburg emperors of Austria. Well into the nineteenth century the heart of Vienna remained a medieval tangle of alleys and streets surrounded by walls, glacis (i.e., a defensive slope), and a moat. These had proved their worth over the centuries in safeguarding the city from the Ottoman Turks, but in 1848 they failed to protect the emperor from Vienna's own citizens, who rose up and drove his troops from the city. In 1857, with the monarchy safely restored,

Emperor Franz Josef I ordered the demolition of the old fortifications, which had outlived their military usefulness and now lay at the center of a city that had expanded far beyond their bounds in every direction. In their place he ordered the creation of a monumental *Ringstrasse* (ring road) that would follow the path of the old walls, surrounding the *Innere Stadt* (inner city) and the emperor's own Hofburg Palace. As envisioned by Franz Josef and his ministers, the Ringstrasse would be the public face of a modern imperial capital.

Construction of the Ringstrasse and the monumental buildings ranged along its path proved to be a colossal project that lasted for more than half a century. But on the eve of World War I the grand boulevard of Vienna was effectively complete. Shaped like a seven-sided polygon, and over three miles long, it was home to the key political and administrative organs of the state and the city, as well as the most prestigious institutions of high culture. One segment of the polygon, known as the Universitätsring, was lined with the Imperial Parliament, the Rathaus (City Hall), the city theater, and—of course—the University of Vienna. Another, known as the Burgring, sported the Palace of Justice and the Natural History Museum. The Opera House looked down on the Opernring, and the Vienna Concert Hall, the Museum of Applied Arts, and the City Park were ranged, quite naturally, along the Parkring.

In some ways the Ringstrasse was a direct descendant of the geometrical style perfected in Louis XIV's capital. In Vienna as at Versailles, a grand ceremonial boulevard lends precise harmony and order to the organs of the state while highlighting the splendor and power of the monarchy. Yet there is no denying that the geometry of the Ringstrasse is fundamentally at odds with the patterns of Versailles: the geometry of Le Nôtre's garden is strictly linear, pointing from the woods of the Grand Parc and the parterres and bosquets of the Petit Parc directly at the royal palace. It thereby established not only an orderly world, in which each object and person has a proper place, but also a strictly hierarchical one, with the king at its apex. The Ringstrasse, in contrast, is laid out not as a straight line, but as a polygon. This means that while each segment is broad and straight, just like the Allée Royale or Unter den Linden, the overall geometrical pattern is a circle, in which all points are equal to all others.

Instead of a single straight boulevard leading up to the Parliament, or the Palace of Justice, or even the university, all these institutions are ranged side by side along the Ringstrasse, without any one dominating the others. As for the imperial palace, the natural focal point of a design inspired by Versailles, it is not even on the Ringstrasse, but off-center, in the area enclosed by it. And while this may be considered a place of honor, it also puts the emperor's authority in an ambiguous relationship to the key institutions of state and society. There is nothing comparable here to the clear lines of hierarchy and power made manifest in the Sun King's garden. Louis XIV, it hardly needs saying, would never have allowed such ambiguity.

The circular Ringstrasse bespeaks a rational and orderly empire, administered by a progressive government and cultivating the universal values of high art and culture. It does not, however, speak of the strict hierarchy that was a key feature of linear Versailles, not to mention the Habsburg emperors' own gardens at Schönbrunn Palace. This, to be sure, does not make the Ringstrasse democratic, much less egalitarian. The institutions ranged along it were all restricted to an economic, political, and cultural elite that dominated the imperial government in the second half of the nineteenth century. The lines of hierarchy and power, so clear and explicit at Versailles, may have been obscured in Vienna. But the line separating those who belonged in the grand edifices of the Ringstrasse from those who did not was as sharp and clear as ever.[14]

It is perhaps ironic that it was in Paris, the city that Louis XIV had sought to escape, that the contours of Versailles left their most famous and recognizable imprints. And they did so, for the most part, long after the last of the Sun King's heirs had departed the throne of France. The French Revolution of 1789 to 1799 brought a bloody and dramatic end to the "old regime," as the political order of Louis XIV and his successors had become known. Even though the Bourbons were temporarily restored in 1814, and a king ruled France until 1848, it was nevertheless clear to all that the dream of royal absolutism could never be recovered. Yet the regimes that ruled France from the revolution onward were just as intent as the Bourbons on creating a strong state administered by a rational central bureaucracy.

Built over centuries without a master plan, or even a broad vision, old-regime Paris was a giant medieval city of narrow streets and winding alleys, where humble residences and artisan shops sat cheek by jowl with great cathedrals, palaces, and monuments. Louis XIV's descendants were, for the most part, as ambivalent about Paris as the Sun King himself, and did little to spruce up the capital. Their showcase was Versailles, and they saw little reason to attract attention to the great and turbulent city they had left behind.

In 1667 Louis XIV did nonetheless give Paris its first touch of the geometrical style by commissioning Le Nôtre to extend the central axis of the Tuileries gardens an additional mile and a half westward, laying out what would later become the first and grandest of Paris's great boulevards—the Avenue des Champs-Élysées. Le Nôtre's goal, it seems, was to turn Catherine de Medicis's enclosed garden into an open perspectival garden like the one he had created for the Sun King. Like the Allée Royale in Versailles, the avenue emerges from what was then the center of the Tuileries Palace and continues broad and straight into the distance before disappearing at a vanishing point. Over the following two centuries, kings, emperors, republican officials, and even revolutionary dictatorships each saw fit to add to the beauty of Le Nôtre's avenue: Louis XV added a grand plaza named after himself, while the revolutionary authorities changed its name to the Place de la Concorde and replaced the king's equestrian statue with an obelisk. Napoleon I added the Arc de Triomphe du Carrousel to the east of the Tuileries Palace and a grander Arc de Triomphe at the opposite (western) end of the avenue, a project that was completed under Louis-Philippe in 1836. Even the destruction in 1871 of the avenue's point of origin, the Tuileries Palace, did little to dim the aura of the Champs-Élysées.

Yet Paris was not just the Champs-Élysées. Even in the middle of the nineteenth century the city was still, for the most part, the same hodgepodge of local neighborhoods, narrow streets, and winding alleys that Louis XIV was so eager to leave behind. And so it was left to one of the most maligned rulers in modern French history, Emperor Napoleon III, to shape Paris into the capital we know today. The mastermind of the new Paris was Georges-Eugène Haussmann (1809–1891), not a great architect in the tradition of Alberti or Wren but a career bureaucrat, who in the 1850s and '60s served as prefect of the Seine district in Napoleon III's Second Empire. Backed by the irresistible power of a centralized state, Haussmann oversaw the wholesale demolition of centuries-old neighborhoods, complete with their medieval buildings and narrow alleys, wherever they stood in the way of his plans. In their place he created broad, tree-lined boulevards that traversed the city in straight lines, north to south and east to west, and connected its major points. Other than the Champs-Élysées, all the most famous boulevards in the City of Lights—the Boulevard Saint-Michel, the Boulevard Saint-Germain, Boulevard de Sébastopol, and on and on—were created under the baron's iron hand. Where the boulevards intersected Haussmann built monumental squares. Where several intersected, as they did at the Arc de Triomphe, he re-created the famous étoile or "star" so familiar from the geometrical landscape of Versailles.

With his wide, straight boulevards, open plazas, squares, and étoiles, Haussman had brought Versailles to Paris. The great and ancient city could not, of course, be transformed into the precise geometrical landscape of Le Nôtre's gardens, but Haussmann's reconfiguring of the city nevertheless had a dramatic effect. Paris became what it remains today—the epitome of a modern capital. Walking through the city's great boulevards, its squares named for great battles and victorious generals, past civic palaces, universities, and cathedrals, both Parisians and visitors are reminded at every turn that they are at the heart of a powerful and orderly state governed by an efficient and rational bureaucracy.[15]

Order and Empire

It was not long before the impact of Versailles was felt far beyond the confines of Europe, reaching out to every corner of the globe. Between the sixteenth and nineteenth centuries the rival European powers established global empires that covered much of Earth's surface and ruled much of its population. As Spain, Portugal, England, France, Holland, and (later) Germany and the United States established colonies from South America to Southeast Asia, and from India to Australia, they built new neighborhoods and sometimes entire new cities to serve as their colonial headquarters. The rational and hierarchical geometries of Versailles, they soon found, were a natural fit for the colonial context, where small clusters of white Europeans claimed the right to rule over immense territories and native populations many times their own number.

In 1859, as part of a "punitive" campaign against the emperor of Vietnam, a joint French and Spanish force stormed and occupied the strategic town of Saigon, not far from the mouth of the river of the same name. When peace was signed three years later the French had succeeded in wresting most of southern Vietnam from the emperor and claiming it as their own. They called the new colony Cochinchina and in 1867 designated Saigon as its capital. Then, with hardly a pause, they began construction of the key edifices that would mark the city as a center of French power.[16]

First came the governor's mansion. Already in 1863, shortly after the conclusion of the war, a wooden structure was erected to serve as the center of French rule, but it was never intended as anything more than a stopgap measure. Work on the permanent governor's residence and colonial headquarters began five years later, when the admiral-governor of Cochinchina ceremoniously laid down the first stone of what would become the Norodom Palace—named for the king of the nearby French protectorate of Cambodia. The disastrous Franco-Prussian War, and the fall of Napoleon III's Second Empire, slowed construction on the mansion, but in 1873 the governor moved into his new home. And a magnificent home it was. Even a quick look at the design or pictures of the old palace leaves no doubt as to its architectural inspiration: it was built as a variant of a French château from the era of Louis XIV, comparable in design and scale to Vaux-le-Vicomte.

Despite its grand design, which harkened back to an illustrious historical era, the Norodom Palace was ill-starred from the beginning. In 1887, only fourteen years after the first governor settled in, Cochinchina was incorporated into the greater colony of French Indochina, whose capital was in Hanoi, far to the north. Considered too grand for the deputy governor's residence in Saigon, the palace was largely abandoned, serving only for ceremonial purposes for the remaining decades of French colonial rule. When the French finally withdrew from Indochina in 1955 it became the official residence of the president of South Vietnam, but it was severely damaged in 1962 when two air force pilots dropped their bombs on the palace in an attempt to assassinate their own leader. The Norodom, known at the time as Independence Palace, was then demolished, but its modernist replacement proved no more fortunate. It was home and headquarters to Nguyễn Văn Thiệu, the last president of South Vietnam, from 1966 until he was forced to flee his homeland in 1975.

With plans for the governor's residence in place, the colonial authorities in Saigon turned their attention to the other grand edifice essential to any French city: a Catholic church. As with the governor's mansion, a wooden church was built in Saigon in 1863, shortly after peace was signed, to serve both the colonists and local converts. The church was functional enough, but with the sparkling Norodom Palace rising nearby, it could no longer be considered grand enough for a colonial capital. And so in 1877 the local bishop laid the first stone for the Cathédrale Notre-Dame de Saïgon, to be built entirely of red bricks imported from France. Construction proceeded rapidly, and in April of 1880 the cathedral was consecrated and services began. A decade later towers were erected on each side of the nave's main entrance, giving the church a medieval gothic appearance that recalled the great European cathedrals. And if the architecture of the Norodom Palace was meant to evoke memories of Versailles, Notre-Dame de Saïgon, in both name and design, was designed to bring up the image of another iconic edifice from the home country—the cathedral of Notre-Dame de Paris.

The palace and cathedral of Saigon served as brilliant emblems of French civilization and power, both worldly and spiritual, over France's colonial subjects. But it was the grand boulevards leading up to these edifices that put the stamp of French authority over the urban landscape. The smaller of these was the rue Catinat, a leafy avenue leading straight from the riverbank to the front entrance of the cathedral, which, according to a visitor in 1893, "distinguishes itself . . . by European looking houses, with shop fronts and a sparkling lightening."[17] Known today as Đồng Khởi Street, it was reputedly built as a copy of the fashionable rue de la Paix in Paris. Then there was the broad and

FIGURE 25: **Boulevard Norodom leading up to Norodom Palace in early-twentieth-century Saigon**

green Boulevard Charner, known today as Nguyễn Huệ Pedestrian Street, which ran parallel to the rue Catinat and ended at the entrance to the baroque-style City Hall. But by far the grandest avenue in the city was the mile-long Boulevard Norodom, today's Lê Duẩn Street. As the city's main ceremonial artery, the boulevard proceeded in a straight line northeast to southwest, past marine barracks, the apse of the cathedral, and a statue of the statesman Gambetta on to the exact center of the Norodom Palace.[18]

The scheme of a grand boulevard leading up to the palace gates was, of course, a standard feature of French royal châteaux and their imitators. At Versailles, the avenue de Paris led straight from the town to the center of the palace, which served as a barrier between the bustling town and the serene gardens beyond. This was also the case in Saigon: the Norodom Palace separated the grand boulevard from the gardens, which extended for half a mile behind the mansion itself. The Palace, gardens, and grand boulevard of Saigon were all built in conscious imitation of Versailles, and the message imparted to the inhabitants of the colony could have come directly from the court of the Sun King: French Indochina was a prosperous, peaceful, and rationally ordered land under the benevolent but firm rule of the governor in his palace. Any challenge to this divinely ordained hierarchy was a strike not only against the might of the French Empire, but also against the universal laws of geometry.

Utilizing the geometrical patterns of Versailles seemed like a natural choice for French colonial officials who were, after all, the direct heirs of Louis XIV and his ministers. It is more surprising, however, to find the same patterns in the colonies of the great English-speaking powers, Britain and the United States. Britain, after all, had been France's enemy and chief rival for European dominance more or less continuously from the days of Louis XIV to the rise of Germany at the end of the nineteenth century. The English political system, anchored in parliamentary supremacy, grew in conscious opposition to French absolutism, just as the naturalistic aesthetics of the English gardening style were meant to contrast with the regimented geometricism of Versailles. As for the United States, its close alliance with France in its early years was based on a shared enmity toward England and could not mask the fact that its republican constitution was fundamentally inimical to the political order of the old regime. Not coincidentally, American designers and planners tended for the most part to steer clear of the example of Versailles, preferring a Cartesian grid pattern for their cities and an English gardening style for their parks. Yet when it came to their colonies, both England and the United States were happy to follow the French lead, producing some of the most complete examples of Versailles-style city planning to be found anywhere.

By the early twentieth century British authorities had grown weary of their old capital Calcutta, which had served as the headquarters of British India since the 1750s. While the city remained a key commercial hub, it was also awkwardly located at the eastern tip of the vast diamond-shaped subcontinent. Even worse, Calcutta had become the center of a fast-growing independence movement that advocated expelling the British from India. To escape a boycott of British goods and violent attacks against imperial officials, Viceroy Charles Hardinge (1858–1944) and his administration decided to remove themselves inland, to a new city that would be built near the ancient Mughal capital of Delhi. The announcement of the move was made in December 1911 by King George V, who was in Delhi for a gathering of notables known as an Imperial Durbar, which culminated in his coronation as Emperor of India. A few days after the ceremonies, the new king-emperor and his queen laid down the foundation stone of the new imperial capital. It would be called New Delhi.[19]

The job of designing the new capital and its main buildings was first given to Edwin Lutyens (1869–1944), a London-based architect who specialized in country houses in the neoclassical style, descended directly from the teachings of Alberti more than four centuries before. When Hardinge grew alarmed at some of Lutyens's more radical proposals he paired him with Herbert Baker (1862–1946), a colonial architect known for his design of the government buildings in the South African capital of Pretoria. Like Lutyens, Baker too was a committed classicist who believed (following Christopher Wren) that good architecture must have "the attribute of [the] eternal." It follows, he argued in a 1912 editorial to *The Times* of London, that the new capital must be built in a neoclassical style. It was the only style, he insisted, that has "the constructive and geometrical qualities necessary to embody the idea of law and order which has been produced out of chaos by the British Adminstration."[20]

By 1913, thanks to Hardinge's strong backing, Lutyens and Baker's design had been approved, and funds were allocated for the construction of the new city. The outbreak of World War I the following year put the plan on hold for several years, but with the war over and won, work on the new capital resumed at a frenetic pace. Monumental neoclassical buildings went up, and government officials and ministries moved in. Finally, in February of 1931, with the main structures completed and

DELHI

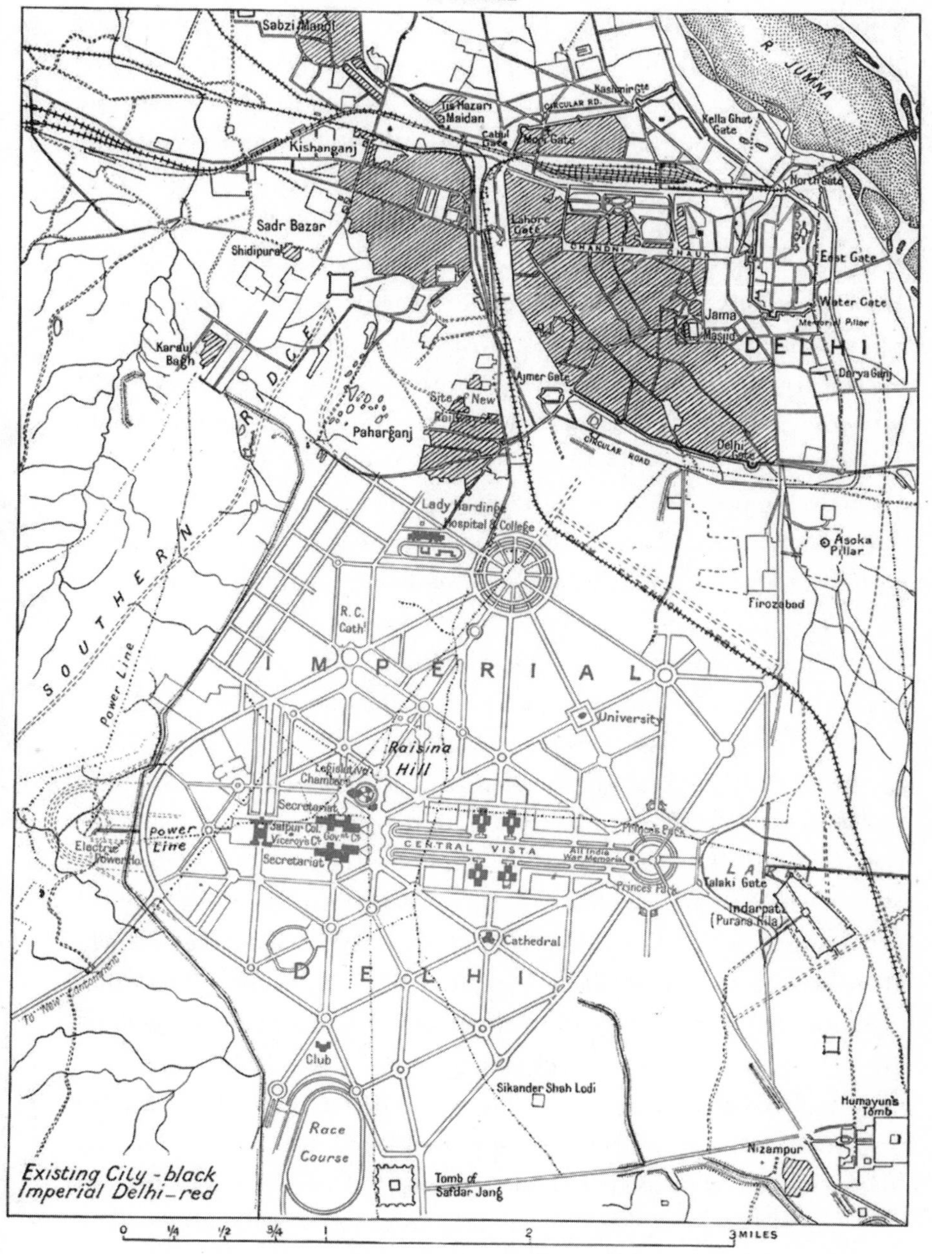

FIGURE 26: Edwin Lutyens and Herbert Baker's plan for imperial Delhi, 1912; the city was later christened New Delhi

the viceroy and his government settled in, the capital was inaugurated with official celebrations that lasted an entire week. It was a glorious display of imperial might and splendor, but the British might have saved themselves the trouble: only sixteen years later British imperial rule crumbled, and the city built to serve the Raj became the gleaming capital of the new independent Republic of India.

When Lutyens and Baker worked out their plan, however, the notion that the end of empire was only a few decades off would have seemed to them patently absurd. Their capital was built to last, and at its heart was Government House—later known as the Viceroy's Palace—seat of the imperial government of India. Hardinge, the viceroy, wrote in 1912 that "the approach to Government House . . . should be a reproduction of Versailles and its gardens," and Lutyens and Baker obliged, creating a center of imperial rule worthy of the Sun King.[21] The palace itself, designed by Lutyens, was a massive edifice in the classical style, complete with a columned façade and a towering dome.[22] And while Hardinge pressed to incorporate local motifs in the design, Lutyens resisted: "I do want," he wrote to Baker in 1912, "old England to stand up and plant her great traditions and good taste where she goes and not pander to sentiment and all this silly Moghul-Hindu stuff."[23]

From the palace gates, extending directly eastward for two miles, was what the original plan referred to as the Central Vista, which later became King's Way (or Kingsway). This was the capital's central axis and grand ceremonial boulevard.[24] In its western part the axis is flanked by imposing twin Secretariat buildings, designed by Baker, which, along with the nearby palace, form the Viceroy's Court. Proceeding eastward, King's Way became a broad open boulevard surrounded by greenery and two parallel avenues to the north and south before finally merging into a massive octagonal plaza at its eastern end.[25] Leading into the center of the plaza, the avenue passed through the All India War Memorial, the triumphal arch known today as India Gate. In a pattern that cannot but bring to mind the famous trivium at the approach to the palace at Versailles, King's Way was flanked to the north and south by twin boulevards that converge with it on either side, forming an arrow pointed westward at the Viceroy's Court. A symmetrical pair of av-

FIGURE 27: Rashtrapati Bhavan (formerly the Viceroy's Palace) as seen from Rajpath (formerly King's Way); the Secretariat buildings are in the foreground

enues also converged on the central axis at its eastern end, forming a trivium with its tip at the War Memorial. These four avenues, converging on the opposite ends of King's Way, also intersected one another, to the north and south of the midpoint between Viceroy's Court and the All India War Memorial. A north–south avenue, which became Queen's Way, passed through both of these nodes, turning them into six-point star plazas and intersecting King's Way at a precise right angle.[26]

These interlocking thoroughfares were the geometrical skeleton of Lutyen's Delhi. Taken together they formed a diamond anchored at Viceroy's Court and the All India War Memorial, with King's Way as the central east–west axis running between them and the midsection of Queen's Way serving as the central north–south axis. Around it, the rest of the city was structured as an elaborate geometrical diagram, roughly symmetrical along its main axis, and composed of rigid triangles that combined to form hexagons, octagons, and more complex figures. Other patterns familiar from the French formal gardening tradition make their appearance at key locations: a giant twelve-spoke rondelle (or star-shaped plaza) dominated the northern reaches of the city in what would become Connaught Place. The main intersections of the grand boulevards formed major plazas, which housed key civic institutions—the Legislative Assembly (today's Indian Parliament), a university, and a cathedral. Finally, in the tradition of Versailles, the central axis extended westward beyond the Viceroy's House, into a large garden at its rear.

There were, to be sure, key differences between the patterns of Lutyens's Delhi and Le Nôtre's Versailles. Whereas in Louis XIV's town and garden all lines converged on the royal palace, the geometries of the Indian capital were more evenly balanced. One end of the central axis was, inevitably, occupied by the British Viceroy's Palace, as one would expect from Hardinge's insistence that it should replicate Versailles. But the other end was anchored at the War Memorial, glorifying not the British colonists, but the native Indians who gave their lives in service of the empire. This geometrical balancing act was a reflection of the imperial policy championed by Hardinge and others that was willing to gradually concede a degree of autonomy (or "responsible government") to the locals as long as they accepted their place in the empire and did not challenge British rule.[27] In Lutyens and Baker's street plan the British and the Indian are both indispensable for the overall harmony and stability of the geometrically ordered city.

In reality, of course, both politically and architecturally, the two ends of King's Way were far from equal. The British never intended to relinquish their imperial ascendancy, and there was nothing in their promises of "responsible government" that would satisfy the aspirations of Indian nationalists for independence. Inevitably, perhaps, this unshakable truth was manifested in the capital's design. Architecture, Baker wrote in his piece in *The Times*, expressed the fundamental truths of British rule; in New Delhi it did so too well.[28]

The western, "British" end of King's Way was the seat of power: it contained the Viceroy's Palace and the two Secretariats, the central executive of the Raj, which the city's planning committee referred to as "the keystone of the rule over the Empire of India."[29] The Indian end of the boulevard, in contrast, contained no government institutions at all, only a monumental arch in honor of the sepoys who sacrificed their lives for the empire. According to Lutyens, its designer, the memorial stood for "duty, discipline, unity, fraternity, loyalty, service, and sacrifice . . . encouraging continued partnership in the established order, and celebrating the ideal and fact of British rule over India."[30] He could hardly have been clearer: like other regimes founded on the supremacy of a white ruling class, the British were happy to acknowledge the courage, or "heart," of their imperial subjects—at least as long as it was manifested in their service. But the "brain" of the operation, that which controlled the levers of power, was to remain safely in British hands.[31]

The ideological differences between the designs of New Delhi and Versailles, it turns out, are largely superficial. Whereas Versailles boldly proclaimed absolute royal supremacy, Lutyens and Baker's plan appears more circumspect, seeming to appeal for a balance between the British and Indian components of the state. Upon closer inspection, however, this balance turns out to be in large part a sham. In Delhi as in Versailles the design is essentially hierarchical and conveys a single message: the ruler, whether king or viceroy, in his palace reigns supreme, secured in his position by the universal, unchallengeable laws of geometry.

For it is, in the end, the laws of geometry that underlay the urban order of New Delhi. The rigid triangles, hexagons, and octagons created a fixed, unalterable, and permanent order that could not be tampered with. The palace and seat of government could not be moved, nor could the War Memorial, for both were held in place by the inflexible geometry of the city streets. The same, in fact, was true for each and every street, plaza, and intersection: none of them could be moved, bent, stretched, or altered in any way; each and every one had to remain for eternity in the place allotted to them by Lutyens and Baker, the city's twin demiurges. Being geometrical, the capital spoke of an orderly Raj and empire, ruled by the universal principles of reason. It spoke of a harmonious world, in which everyone—British and Indian alike—occupied their appropriate place in a single grand scheme, determined by reason and necessity. And it spoke of hierarchy, for each of these positions is strictly ranked on a scale of honor and power, with the British, inevitably, occupying the top rungs. Universal order, harmony, and hierarchy—these were the principles that Lutyens and Baker encoded in the streets of the new capital. And all were founded on the eternal laws of geometry.

Unlike France and Britain, the United States was a relative latecomer to the Great Game of overseas empire. A young nation with continental ambitions, it was preoccupied throughout the nineteenth century with its own western frontier and had little time or resources to devote to imperial adventures. And that was not all: the United States, after all, was a republic, founded on the principle that "all men are created equal." And even though in practice Americans often failed to live up to that dictum, they were nevertheless uneasy about the prospect of invading faraway lands and dominating their people through sheer military might. But when in 1898 a brief war with Spain left the United States in possession of Puerto Rico, Guam, and the Philippines, such scruples were set aside. Suddenly Americans had their own overseas domains, and they were eager to learn from the experience of the senior imperial powers. Not least among the lessons they absorbed from their French and British mentors was the importance of establishing a colonial capital worthy of their imperial standing. And so, in the Philippines, the Americans set out to build one.

At the turn of the twentieth century Daniel H. Burnham (1846–1912) was far and away the most famous and influential of American architects. He had earned his reputation a decade before by serving as chief designer of the grounds for the World's Columbian Exposition in Chicago. The temporary White City he built for the exposition on the shores of Lake Michigan, complete with grand boulevards and classically themed buildings, was acclaimed as a marvel of functionality and beauty. Then, in 1901, he followed this triumph as chairman of the U.S. Senate Park Commission, charged with making Washington, D.C., into a capital worthy of America's rising place in the world. His plan for restoring Pierre L'Enfant's Versailles-inspired plan from 1791 and adapting it to modern conditions has shaped the city to this day.[32]

So things stood in 1904, when Burnham received a query from his young friend William Cameron Forbes (1870–1959), who had recently been appointed Commissioner of Commerce and Police in the Philippines. Forbes was looking for an architect to redesign the Philippine capital of Manila and wanted to consult Burnham on possible candidates. Burnham, looking for new challenges, suggested himself; Forbes, embarrassed that he had not thought to make the offer, immediately accepted. A few short weeks later Burnham was on his way to the Philippines, arriving in Manila on December 7. He would spend the next six weeks on the island of Luzon, dividing his time between Manila and the mountain town of Baguio, which the American authorities were hoping to transform into a government retreat for the summer months. Returning to the United States in February of 1905, he began to work out his plan to make the Philippine capital "the Pearl of the Orient."[33]

Burnham, quite to his surprise, was enchanted with Manila. "[The] dive into the orient has been like a dream," he wrote to his friend (and future biographer) Charles Moore after his return. "It surprises me to find how much this trip has modified my views, not only regarding the extreme East, but regarding ourselves and all our European precedents."[34] The city's setting, he wrote in his 1906 report, is "unparalleled and priceless," and the tropical Spanish architecture "unusually pleasing."[35] And yet when he actually laid down his plans for the Philippine capital, Burnham seemed to have put all that aside. The local style may have been charming, and undeniably practical for the tropical conditions, but it was not appropriate or suitable for a colonial capital. For that, his plan made clear, only the grand classical style familiar from Saigon and New Delhi was acceptable—the same style Burnham had perfected in Chicago and Washington, which was inspired, ultimately, by Versailles.

FIGURE 28: **Daniel H. Burnham's plan for Manila, 1905; the map is facing east**

Just as Lutyens's Delhi was built around a core group of government buildings, so it was with Burnham's plan for Manila. The key structure here was not a governor's palace, but a domed and columned capitol, modeled on the U.S. Capitol in Washington and intended to house the Philippine Legislature. Clustered around it in a "hollow square, opening out to the westward" would be a Hall of Justice, library, museum, exposition buildings, and a post office.[36] Together these edifices would form what was effectively a national mall—a broad parklike avenue leading from the center of the capitol to the waterfront, flanked on both sides by government buildings. From the eastern face of the capitol, on the opposite side from the mall, linear avenues would spread through the city. The practical advantage of this arrangement, Burnham explained in his report, is that "the center of governmental activity should be readily accessible from all sides." The symbolic implications, however, were even greater, because, thanks to the radial plan, "every section of the Capital City should look with deference toward the symbol of the Nation's power."[37]

The rest of the city, in Burnham's plan, follows a strict geometrical pattern, not very different from Lutyens's Delhi. Broad and straight boulevards reach out to every corner of the city, where they intersect with similar boulevards at precise angles. The result is a web of interlocking triangles covering the entire expanse of Manila, forming squares, hexagons, and octagons, and meeting at their tips to form monumental star-shaped plazas. The overall design is consequently rigid and immovable, as was the case in Delhi. Every avenue, street, and alley has its precisely defined place in the grand geometrical design, from which it cannot stray. And all of them look up from their location "with deference" toward the city's capitol, the seat of power and legitimacy.

All of these features from Burnham's plan should sound familiar. The idea of establishing new centers of power outside the old city walls guided the creation of the Ringstrasse in Vienna and the new imperial capital outside of Mughal Delhi. The pattern of the wide straight boulevard leading up to the seat of power, the cluster of classically designed government edifices, the grand avenues radiating from them (or converging toward them) shapes the Unter den Linden in Berlin, the Admiralty building in St. Petersburg, Norodom Palace and Boulevard in Saigon, King's Way and the Viceroy's Court in New Delhi. All of these features, furthermore, can be seen in Versailles, the original geometric capital outside the old city walls and the inspiration for all that followed. For all of them, European and colonial cities alike, are the descendants of Louis XIV's capital.

From New Delhi to Manila, and from Rabat to Canberra, the geometries of Versailles were the primary choice of colonial urban planners. Part of the reason was undoubtedly practical: imposing a geometrical street plan on European cities, with their winding streets and ancient residences and churches, required a wholesale destruction of property that was extraordinarily difficult in the best of times and simply unfeasible in most others. Emperor Franz Josef, to be sure, succeeded in transforming the old walls and moat of Vienna into the sparkling Ringstrasse, and Baron Haussmann, backed by the imperial authority of Napoleon III, demolished entire Parisian neighborhoods to create his network of elegant boulevards. Yet those were rare and, on the whole, partial successes that left most parts of the cities untouched. Charles II had set his eyes higher when he tried to exploit the devastation of the Great Fire to build his own Versailles in London, only to ultimately concede defeat when faced with an impenetrable web of legal claims.

But if European monarchs were hamstrung at home by local interests and jealously-guarded laws and traditions, in the colonies things were different. Backed by overwhelming military might, unconstrained by local customs and interests, and often indifferent to the suffering of the population, colonial administrators ruled their domains with near-godlike powers. If they ordered a new avenue built, the workers set out with their shovels; if they ordered a neighborhood destroyed, it would be gone within weeks; and if they dreamed of a brilliant new city where only stony hills could presently be seen, it would not be long before grand edifices would rise up from the barren rock. Consequently, a plan to create a Versailles-inspired city in Europe was unlikely ever to leave the drawing board; but in the colonies it could and did become a reality.

There was more, however, to imperial architects' affinity for the geometrical style than simply the promise of executing complex, ambitious plans in a colonial context. For in the eyes of imperial advocates and colonial administrators, there was no more powerful expression of the ideals and promise of empire than a geometrically ordered city. Empire-building, at its core, was nothing more or less than the forcible occupation of faraway territories and the subjugation of their people. But advocates of empire and the colonists (at least those who cared to justify their actions) did not consider themselves brutal occupiers, but rather idealists on a mission. The Spanish conquistadores and Portuguese settlers of the sixteenth century, for example, declared themselves honor bound to spread the true faith to every corner of the world, even at the point of a sword. Many of them undoubtedly believed it, too.

Northern European imperialists took a different stance: it was not the gift of faith that Europeans were bringing to the subject lands and peoples, but the gift of reason. Taking root at the height of the age of Enlightenment, then spreading rapidly in the nineteenth century, the outlook of the French and British empires was shaped by a heady belief in the benefits of rationality and science. To imperial apologists the colonists brought the light of reason and technological progress to benighted peoples, and law and order to lands where strife and disorder ruled. Far from being brutal oppressors, these emissaries of empire were the bearers of progress, sent out to civilize the backward lands and peoples of the world.

For those who believed in the civilizing mission of empire, the choice of a geometrical plan for colonial cities was an obvious one. Like the geometries of Versailles, the city plans of New Delhi or Manila were emblems of good government founded on universal reason. The broad, straight, and open boulevards cut through the winding back alleys of indigenous towns like the clear light of reason slicing through ancient superstition.[38] The rigid patterns of triangles, hexagons, and octagons placed the boulevards, intersections, and plazas in a fixed geometrical design in which each supported all others. Each had its place and none could be moved—a valuable lesson for the imperial subjects.

Most important, the world of these capitals of empire was hierarchical to the core. In Saigon, New Delhi, Manila, and elsewhere, the great avenues led unswervingly to the centers of imperial power, which ruled majestically over the city and its population. The message imparted was double-edged, but no less powerful for it. On the one hand, the geometrical design of the city streets was clear evidence that the white colonists had brought order and reason to superstitious, chaotic lands; on the other, those same patterns buttressed the rule of the colonists and made it into an irrevocable part of a rational universal order. The cities, in other words, demonstrated the civilizing mission of empire while securing its timeless rule. No message could be more welcome to imperial authorities, both in the colonies and at home.

The absolute monarchy of the Bourbons is long gone, as are the kingdoms of their royal rivals who sought to emulate the Sun King. The mighty global empires that followed in their wake and inscribed their rule onto colonial cities have also crumbled into dust. And yet in one city the geometrical legacy of Versailles continues still, not as the fossilized record of a bygone era, but as a living force that still shapes social relations and political order. Paradoxically, this city was built not by autocratic monarchs who sought to rival Louis XIV or power-hungry imperialists intent upon firming up their rule. It was, rather, founded by republican revolutionaries, men who detested everything that the Sun King stood for but like him believed that geometry would buttress the ideals they held sacred. This city is, of course, Washington, D.C., the capital of the largest and mightiest democracy on Earth.

That the legacy of one of history's greatest absolute monarch lives on in the capital of the world's greatest republic is due to the creative imagination, perhaps even the genius, of a single man. He was, not coincidentally, a Frenchman who spent his youth in the courts of the Bourbon kings but made his home in the democratic United States. So unyielding was he in his commitment to applying the geometrical style of Versailles to the design of the republican capital that he brought upon himself the wrath of the most powerful people in his adopted homeland. His obsession destroyed his promising career, but the city he envisioned survived. His name was Pierre Charles L'Enfant.

7.

THE EUCLIDEAN REPUBLIC

The Prickly Designer

It was early April in the year 1791, and Major Pierre Charles L'Enfant (1754–1825), architect and surveyor, was not pleased. Only a few weeks had passed since he had been named the designer of the future capital of the United States of America, and he was spending his days surveying the grounds where the city was to rise and hatching a magnificent urban plan, worthy of the first city of the republic. Yet now, a mere month after securing the coveted appointment, he had been handed a sheet of paper with a crude design of the city as it was to be. It was not what he had in mind.

As the design was sent to him by no less a personage than President George Washington, caution and courtesy dictated that he should

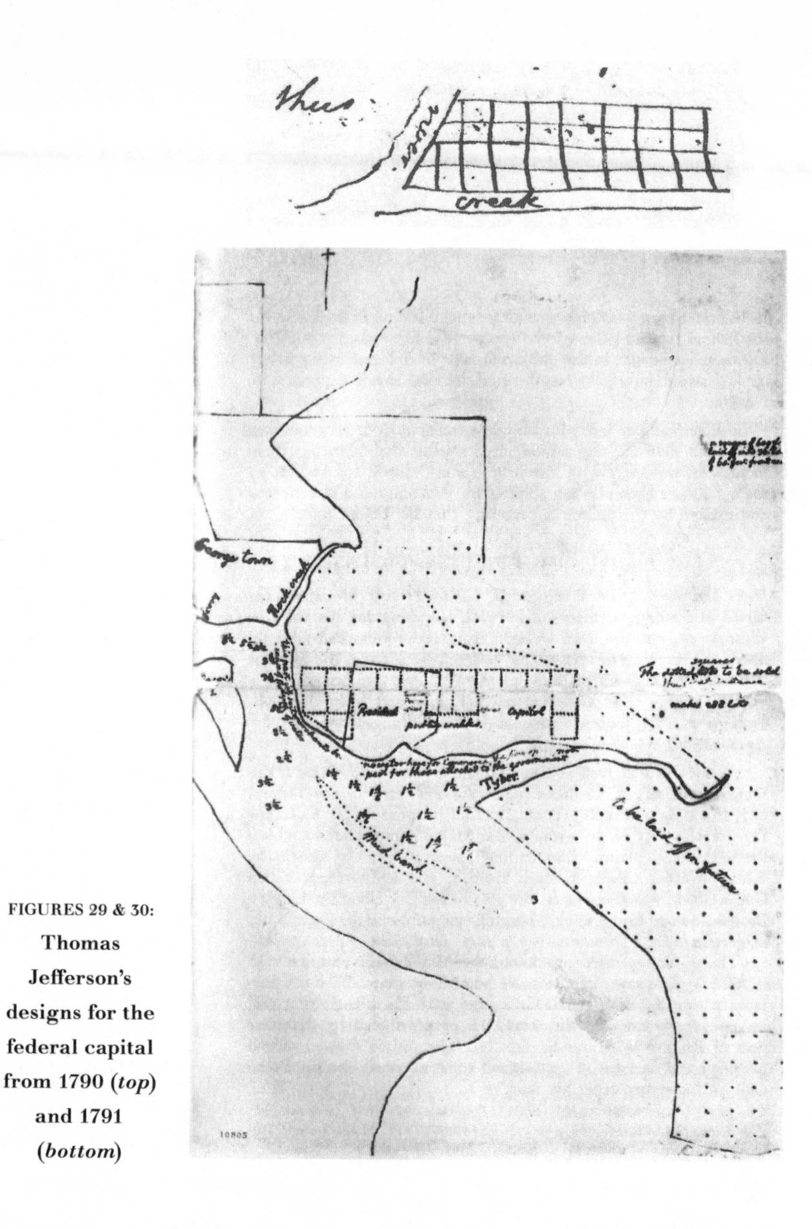

FIGURES 29 & 30: **Thomas Jefferson's designs for the federal capital from 1790 (*top*) and 1791 (*bottom*)**

treat it with the respect due to the nation's first citizen. But L'Enfant's disgust overwhelmed both good sense and good manners. The plan, he told the president, was a disaster, one that would render his role as designer entirely worthless. "A plan of this sort," he fumed, "must be defective, and it never would answer for any of the spots proposed for the

Federal City . . . it would absolutely annihilate every [one] of the advantages enumerated and . . . alone injure the success of the undertaking."[1]

Perhaps L'Enfant believed he could afford to be so blatantly outspoken even when addressing the most powerful man in the land. After all, though a Frenchman, his American bona fides were impeccable: having volunteered in 1777 to serve in the American cause, he had soldiered gallantly in Washington's Continental Army until he was injured in the siege of Savannah in 1779. He then spent the remainder of the Revolutionary War as an engineer on Washington's staff, where the general took a liking to the French firebrand. After the war, boosted by the patronage of the nation's greatest hero, L'Enfant set up a successful surveying and engineering office in New York City. In 1788 his good connections landed him a commission to transform New York's old City Hall into the imposing Federal Hall, where Congress met while more permanent plans were made for housing the national government. By 1789, while Congress was still hotly debating the proper site of the future permanent capital, L'Enfant was already offering himself to Washington as the man best suited to design it.[2]

And yet, for all his success and connections, L'Enfant's high-handed rejection of the preliminary design for the new capital must have seemed to contemporaries more than a tad rash. For the author of these plans, as L'Enfant well knew, was Secretary of State Thomas Jefferson, a man second only to Washington himself in the esteem of his countrymen. He was, of course, the chief author of the Declaration of Independence, though this fact may not have been widely known at the time.[3] He had also been, successively, the wartime governor of Virginia, a leading member of the Continental Congress in the early years of peace, and, until recently, minister to France, the United States's greatest ally on the international stage. Following the ratification of the Constitution in 1788, he was recalled from Paris to serve in Washington's newly formed cabinet.

All this should have given L'Enfant pause before he declared Jefferson's brainchild a "defective" plan that would "annihilate" and "injure" the future capital. Early Americans may have believed that "all men are created equal" in principle, but they were also keenly aware that that was not so in practice, and a man like Jefferson, a gentleman landowner and member of the Virginia elite, vastly outranked the émigré hireling L'Enfant. Add to this the fact that as secretary of state Jefferson was one of the most powerful officials in the country, and one can get a sense of just how reckless the architect was in so rudely (if indirectly) dismissing his better's suggestion.

Yet there was even more reason for L'Enfant to tread carefully in this particular case. More than any man alive, it was Jefferson who was responsible for the fact that the new city was to be designed and built at all—and that L'Enfant, for that matter, had a job. While Article I of the Constitution had decreed that Congress establish a "seat of Government" on land ceded by the states, it left open the question of where that seat would be located. And for several years, until Jefferson's intervention, it appeared likely that no location would be agreed upon, no land would be ceded, and no capital would be built.

Candidates were not lacking, as towns and cities up and down the eastern seaboard, eager to raise their profiles (not to mention their real estate values), sent letters to Congress offering themselves as suitable homes for the federal government. Some towns were modest, like Kingston, New York, and Annapolis, Maryland, but others were the great metropolises of the republic: New York City, Baltimore, and Philadelphia. Yet despite long debates and several votes on the issue, Congress in 1790 seemed no closer to picking a site for the permanent capital than it had been when the Constitution was ratified two years before.[4]

The reason was politics: "the question of residence," as it became known, had fallen victim to the regional rivalries between the states that composed the newly founded union. Since everyone could see the advantages of having the national capital located on their own turf, no one was willing to concede the honor to their rivals. The northern states preferred a site in New York, close to the bustling port city that was already on its way to becoming the commercial capital of the nation. Mid-Atlantic states supported a site in Pennsylvania, extolling its central location and historical credentials: it was, after all, in Philadelphia that the Continental Congress had met in 1776 to declare independence from Great Britain, and there that state delegates had met eleven years later to draft the Constitution. Southerners, meanwhile, lined up behind Virginia, the most populous and wealthiest of the states and—in its own view, at least—the union's natural leader. To Virginians a site on the Potomac River, the great waterway connecting the Chesapeake Bay to the Blue Ridge Mountains, was the obvious choice for a nation looking to expand westward.

Leading the charge in Congress for the Potomac site was Jefferson's friend and collaborator James Madison. During the Constitutional Convention and the struggle over ratification, Madison was an outspoken advocate of a strong national government, working to soothe the fears of skeptical southerners who worried that a centralized federal government would be just as oppressive as King George was in his day. That Virginia and its neighboring states ultimately overcame their misgivings and ratified the Constitution was in large part Madison's doing. But with the Constitution ratified in 1788 and the Bill of Rights passed in 1789, Madison returned to his roots, championing the cause of his native state. The upper reaches of the Potomac, he argued in a speech to Congress, where Conococheague Creek empties into the river, are a gateway to the Ohio Valley and through it to the great Mississippi.[5]

Unfortunately for Madison, his eloquence was wasted on his northern brethren. If the Potomac River itself seemed excessively remote to northerners, the mention of Conococheague Creek settled their view on the matter. "Enquiries will be made," observed one Massachusetts congressman, "where in the name of common sense is Connogochque?"[6] No one outside of Virginia seemed to know, let alone consider it a suitable site for the national capital. Madison's advocacy of the Potomac site appeared to New Englanders as nothing but a crude power grab by the state of Virginia, which in any case wielded far too much power in the union. And as long as northern delegates to Congress were determined to block it, the Potomac proposal was doomed.

It all came to a head in the summer of 1790, when negotiations over the federal capital had reached a deadlock. The northern states were willing to support a Pennsylvania site near Philadelphia, but firmly rejected the Potomac. The Virginia delegation, meanwhile, though frustrated in their efforts to secure the selection of their favored site, proved highly effective in scuttling congressional approval for any other site. It appeared likely that for the foreseeable future Congress would continue meeting in L'Enfant's elegant Federal Hall in the temporary capital of New York. Finding a permanent seat for the federal government, as the Constitution required, seemed beyond reach.

It was at this critical point, when the debate over the national capital was at a stalemate, that Thomas Jefferson entered the fray. Having spent five eventful years in Paris, Jefferson followed the great debates surrounding the adoption and ratification of the Constitution from afar, aided by a constant stream of letters from Madison. Like many Virginians of his class, Jefferson was concerned about the broad powers granted to the federal government in the Constitution to override the sovereignty of the states. But as the newly appointed secretary of state in Washington's administration, he was determined to make sure that the new government would take root and not be rendered impotent by state rivalries. Establishing a national capital, he realized, where a strong federal government could operate independently of local and regional interests, was essential if the union were to survive and thrive. And yet it was precisely those parochial interests that now prevented the capital's creation. What was one to do?

Jefferson thought he had the answer. Idealist though he was, Jefferson was also a shrewd negotiator and a master of the give-and-take of political life. To get what you want, he knew, you must concede something the other side wants just as badly, and the way to break the stalemate was therefore to strike a bargain between North and South that would serve the interests of both. The challenge was to find an issue that the northern states valued as much as the southern states cared about the Potomac site for the federal city.

One day, in the middle of June, 1790, Jefferson saw his opportunity. As he recalled years later, he was waiting outside the president's office in New York when he encountered his colleague the secretary of the treasury, Alexander Hamilton. Usually a dashing figure bursting with energy, Hamilton was not himself that day. To Jefferson he appeared "sombre, haggard, and dejected beyond comparison," his normally resplendent dress "uncouth and neglected." Weighing Hamilton down was the stalemate over his plan for the recovery of public credit, which he had submitted to Congress back in January. Some northern states, Hamilton explained to Jefferson, had borrowed heavily during the Revolutionary War to finance the fight and now stood on the verge of bankruptcy.[7]

The obvious solution, to Hamilton's mind, was to have the newly formed federal government assume the obligations of the states. This was just, he explained to Jefferson, because the northern states had accrued their debts fighting for the common cause, and abandoning them in their time of need would be a vile act of betrayal. It was also essential, because the outstanding debts were undermining the international standing of the union as a whole. Only a consolidated national debt backed by the combined resources of all the states could restore the republic's credit among foreign lenders.[8] To Hamilton this seemed simple common sense: having joined forces with Madison to pass and ratify the Constitution, he was now working hard to turn the elegant phrases of that document into an institutional reality by creating a powerful federal government. The assumption of the debt was self-evidently a key part of establishing such a central authority, which he considered nothing more than putting the Constitution into effect.

Hamilton may have expected some opposition from the Anti-Federalists, who had fought tooth and nail against the Constitution and were now continuing their struggle against its implementation. What he did not expect was that Madison, his former brother in arms in the Constitutional struggle, would rise up and lead the opposition. Yet so it was, and as long as Madison and his allies were opposed to the assumption bill, the plan for the recovery of the public debt was doomed. Hamilton, who never doubted that commerce and finance were the keys to the republic's future greatness, was in despair. If his plan was not approved, he told Jefferson, he would resign, and the northern states might withdraw from the union.

The reasons for Madison's opposition to the assumption bill were complex. Partly it was a matter of self-interest, since Virginia had already paid off its own debts, and its leaders saw little reason why the state should now assume a new debt burden on behalf of its less fortunate neighbors. But there was more to it than that: in Hamilton's plan Madison recognized the twin specters of an overly powerful central government and a union dominated by urban financial interests. Consequently, whereas Hamilton considered the bill a simple implementation of the Constitution, Madison saw it as a breach of both its letter and spirit. If Hamilton had his way, Madison worried, the nation could fall prey to the corrupting influence of what Jefferson called "that Speculating phalanx."[9]

And so it was that as spring turned to summer in 1790 Madison and Hamilton, so recently collaborators in the great fight to ratify the Constitution, faced each other on opposite sides of an ideological divide. Each led a faction strong enough to block the designs of the other yet was helpless to realize its own. It was the first instance of congressional gridlock in U.S. history, though—as any observer of twenty-first century American politics is sure to note—far from the last. Though Congress, Jefferson wrote in a passage that could have been written today, "met every day, little or nothing could be done from mutual distrust and antipathy."[10]

But where most saw the struggles over the federal capital and the national debt as compounding a growing rift between North and South, Jefferson perceived a chance to bridge it. As he tells it, he invited Hamilton and Madison to dine in his rooms and brokered a compromise: assumption of the debt in return for a capital on the Potomac. Madison promised to moderate his opposition to Hamilton's plan, thereby allowing its passage; Hamilton, for his part, promised to deliver the support of both the New York and the Pennsylvania delegations for the Potomac plan, thereby ensuring that it would prevail. To placate the Pennsylvanians, who believed they had the strongest claim for a federal site, it was agreed that while construction was proceeding on the Potomac, the government would temporarily be housed in Philadelphia.[11]

It wasn't easy, but in the end Jefferson's compromise carried the day in Congress. The vote margins were slim, and the opposition to both the Residence Bill (on the capital) and the Assumption Bill (on the debt) remained strong. And yet, the gridlock had been broken and work could now begin on the new national capital located along the banks of the Potomac. If L'Enfant was to realize his dreams of designing this new American metropolis, he had Thomas Jefferson to thank for it.

Gratitude alone should have induced L'Enfant to stay on the best of terms with Jefferson. But if that proved insufficient, he would have done well to consider that, next to the president himself, Jefferson was the cabinet member most closely engaged in the building of the city on the Potomac. Working closely with Washington, Jefferson had assisted in selecting a suitable site, proposed the size and shape of the federal district, and negotiated with local owners over the sale of their land. For the next three years he kept a close watch over everything from the city's streets to its public buildings and monuments, and from the sale of lots to the style of its architecture.[12] Jefferson knew that the final word about the capital lay with President Washington, who from the beginning was closely identified with the city that would bear his name. But he had every reason to expect that his proposals would be treated very seriously indeed, and with the respect due to his position.

And yet L'Enfant would not. Plans such as Jefferson's, he wrote Washington, "however answerable they may be on paper or seducing as they may be on first aspect to the eyes . . . become at last tiresome and insipid." Nor was Jefferson's design simply the unfortunate result of lack of cultivation and sophistication, excusable perhaps in a new, rustic land. It was, rather, "in its origin, but a mean continuance of some cool imagination wanting a sense of the real grand and truly beautiful."[13] Defective, injurious, tiresome, insipid—L'Enfant seemed to be reaching for the right words to express the depth of his contempt for Jefferson's design. The secretary of state, it seemed to L'Enfant, simply could not fathom the beauty and grandeur he had in store for the new capital.

That L'Enfant paid a price for his insolence is hardly in doubt, though the full consequences for his future took some time to unfold. Initially it seemed that the plainspoken Frenchman would succeed in protecting both his plan and his career. Washington almost certainly showed Jefferson L'Enfant's offending letter, but the secretary chose not to press the matter. When he wrote to L'Enfant a week later, his letter—unlike the architect's—was a model of civility. "I am happy that the

President has left the planning of the town in such good hands," he complimented L'Enfant without a trace of irony, adding that he has "no doubt it will be done to the general satisfaction." As to his own ideas for the design of the capital, he would push them no further. "Having communicated to the President . . . such general ideas on the subject of the Town, as occurred to me," he wrote, "I make no doubt that, in explaining himself to you on the subject, he has interwoven with his own ideas, such of mine as he approved."[14] It was therefore pointless, he continued, to confuse the issue by communicating his ideas directly to L'Enfant. It was up to Washington, he suggested diplomatically, not himself, to rule on the city's design.

There was, in all likelihood, more to Jefferson's restraint than merely his trademark civility. For although Congress had approved placing the capital on the Potomac, critics of the choice were still looking for every opportunity to reverse it. A public feud between the secretary of state and the newly appointed city architect might doom not only the architect's career, but also the entire effort to build a capital on the Potomac. Furthermore, Washington's support made the architect untouchable, even to Jefferson. And so, for the moment, L'Enfant survived his brush with Jefferson. But he had acquired a powerful enemy, one, furthermore, who was profoundly interested and deeply involved in the construction of the capital. Over the course of the following year, as L'Enfant's relationship with Washington frayed and his position grew tenuous, Jefferson's understated though implacable hostility came increasingly into play.

The first signs of trouble for L'Enfant came in August of that year, when he displayed his plan for the capital in Philadelphia. Jefferson, along with Madison, closely inspected the design and presented a list of skeptical and unflattering "queries" to Washington. The plan, they reported, was too grand, too expensive, and ill-suited to both the resources and the spirit of the young republic. When Washington responded to the critique by trying to rein in the ambitious design, L'Enfant persisted. Instead of making the changes, he dug in his heels, and, as if to make his point, ordered the demolition of a house that he claimed stood in the path of a projected boulevard. The house was quickly demolished, but L'Enfant's own position also suffered a fatal blow. The three commissioners appointed to oversee the creation of the city, who were nominally his superiors, insisted that they had not been consulted, and that L'Enfant had no authority to take such unilateral action. Jefferson, not unpleased with L'Enfant's troubles, supported them all the way.

When L'Enfant was ultimately dismissed in March of 1792, less than a year after his appointment, it was Jefferson who struck the final blow: "It having been found impracticable to employ Major L'Enfant in that degree of subordination which was lawful and proper," he wrote the commissioners, "he has been notified that his services were at an end."[15] And that was that. Though famously urbane and averse to open conflict, Jefferson was not an enemy to trifle with. And though he lived for another thirty-three years, L'Enfant died in poverty, never having received a public commission again.[16]

The Constitution in Stone

L'Enfant's reaction to Jefferson's proposal was not only rash, but also entirely out of character. Throughout his time in America the Frenchman had shown himself a master at ingratiating himself with his superiors, gaining the friendship and patronage of the greatest men in the republic. And yet, when presented with nothing more than a rough sketch suggesting the outlines of the future city, his rage and indignation were such that he threw it all away. What was it, then, about Jefferson's plan that set L'Enfant on his path of self-destruction? What was it that so outraged him that, defying both civility and self-interest, led him to so recklessly attack Washington's secretary of state?

To make sense of this seeming madness we need to look back at his personal roots, which shaped both his philosophy and his designs. For L'Enfant, without question, was a child of old-regime France. Born in Paris in 1754, he was the son of Pierre L'Enfant (1704–1787), a successful court painter to Louis XV who specialized in dramatic battle scenes. Throughout his childhood and youth L'Enfant was at home in the great palaces and parks of the Bourbon monarchs. He walked the paths of Versailles and the Tuileries, the elegant Palais Royale and the

Champs-Élysées, imbibing their strict geometrical style. At the age of seventeen he entered the Parisian Académie de Peinture et Sculpture, and he was no doubt expected to follow in his father's footsteps and enter the royal service. His court career, however, was cut short in 1777, when he joined a group of volunteers recruited by the dramatist Beaumarchais and sailed to America to fight for the cause of liberty.[17]

It's hard to know what it was that lured L'Enfant across the Atlantic: it may have been belief in the justice of the colonists' cause, or a thirst for glory, or a sense that his professional prospects at home were limited. Or, as his later career suggests, it may have been all three. For once in America L'Enfant made clear his devotion to American freedom and sought out every opportunity to face the enemy in open combat. Grievously wounded in an attack on British earthworks at Savannah, he was not yet recovered when he was taken prisoner while fighting in the defense of Charleston. Following his release in a prisoner exchange, he earned high praise from Washington and was promoted to Major of Engineers by an act of Congress.[18]

Yet it was not his gallantry under fire or his engineering prowess that secured L'Enfant's reputation in America. It was, rather, that he was the man who introduced the French formal style of design to the United States. Hints of L'Enfant's ability to translate his Parisian training to the American setting were evident early on, when in the bitter winter of 1777–1778 at Valley Forge he was asked to paint a portrait of General Washington, presumably in the martial style perfected by his father. In Philadelphia four years later the French minister to the republic charged him with designing and staging a celebration to mark the birth of the dauphin, son of Louis XVI and Marie Antoinette. In the best tradition of royal celebrations at Versailles and the Tuileries, L'Enfant created a garden pavilion with dramatic representation of French and American national symbols. Six years after that, commissioned to transform New York's City Hall into a suitable temporary home for Congress, he did so in the somber neoclassical style typical of the public buildings of European capitals.[19]

The scope of L'Enfant's ambitions and his commitment to the French style, however, did not reveal itself fully until 1791, when he presented

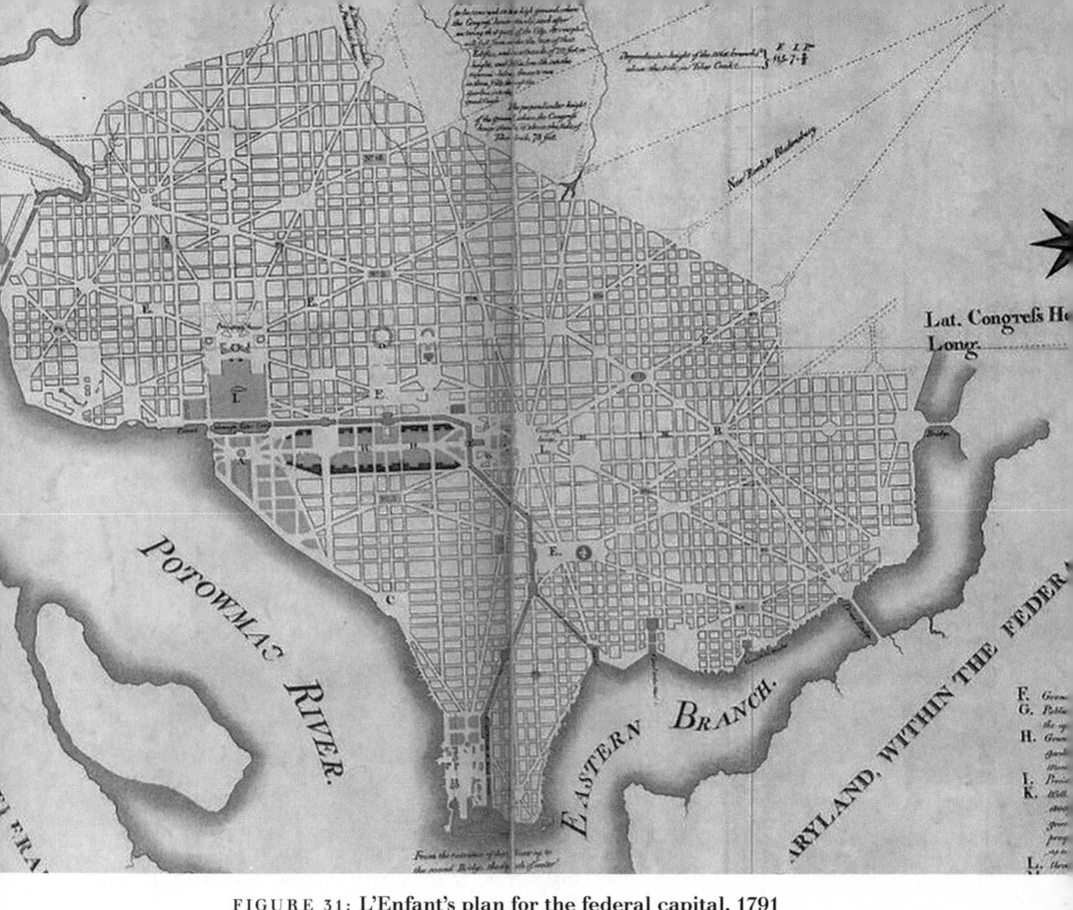

FIGURE 31: **L'Enfant's plan for the federal capital, 1791**

his plan for the future capital of the United States. The first draft of the design, which has not survived, was presented to Washington in June of that year to allow the president to comment and make changes as he saw fit. A revised plan was then made available that August for inspection by Congress in Philadelphia. This is the version that Jefferson and Madison commented on, and it is the one referred to today as the L'Enfant Plan.[20]

And what a plan it was. Bounded by Georgetown to the west, the Eastern Branch (known today as the Anacostia River) to the southeast, and the Potomac to the southwest, it was a city on a scale never seen or even imagined in colonial America. Stretching more than four miles from east to west and nearly as far from north to south, and covering roughly six thousand acres, it was about six times the size of Philadelphia, the largest city in the union. At a time when the population of the

United States was overwhelmingly rural, and no city had more than 30,000 residents, L'Enfant was proposing a city of literally hundreds of thousands. It was a stunningly ambitious plan for a nation of four million that was still recovering from the ravages of war.[21] Yet L'Enfant was far from apologetic: the city, he wrote to Washington, must be "laid out on a dimension proportional to the greatness which . . . the Capital of a powerful Empire ought to manifest."[22]

The sheer magnitude of the plan is enough to bring to mind Le Nôtre's creations in Versailles and Paris, in which the scale of the undertaking was designed to awe all visitors and impress them with the irresistible might of the monarchy. But even more than the size of the city, it was its street plan that echoed the royal gardens and cities of France. The layout of L'Enfant's city would be defined by two dominant sites, each representing a key component of the republic's government. One was the congressional Federal House, to be located at the top of the highest hill in the region, known to L'Enfant as Jenkin's Hill and to later generations as Capitol Hill. In L'Enfant's words, it was "a pedestal waiting for a superstructure."[23] The other site was the President's Palace, farther to the west, about a third of the way toward Georgetown.

From each of these two nodes extended broad tree-lined avenues, 160 feet wide and arrow-straight, radiating in all directions. A full twelve such avenues, three from each of the cardinal points of the compass, converged on the Federal House, and seven converged on the President's Palace. No one who had been to Versailles would fail to recognize the inspiration for L'Enfant's design in the trivium of grand boulevards converging on Louis XIV's palace and emerging again into the gardens, and the famous star-shaped rondelles of the Grand Parc. The broad avenue connecting the two poles of power (known to us as Pennsylvania Avenue) extended beyond them, forming a central axis to the city from the Anacostia River in the southeast to Georgetown in the northwest. Two long avenues ran parallel to the central axis to the north (Massachusetts Avenue) and south of it (Virginia Avenue), intersecting with those emanating from the centers and forming star-shaped plazas along the way.[24]

The formal centerpiece of the design was what L'Enfant called a 400-foot-wide "Grand Avenue," recognizable today as the National Mall. Flanked by broad gardens on each side, the avenue stretches westward from the foot of Jenkin's Hill and ends at an equestrian statue of George Washington near the river, directly south of the President's Palace. It was a republican incarnation of Versailles's Allée Royale, in which Congress occupies the position of the royal palace and George Washington takes the place of the Sun God rising from the waves in the Fountain of Apollo. Washington's statue was also the endpoint of another formal garden, the President's Park, sloping straight down from the President's Palace and intersecting the Grand Avenue at a precise right angle.[25]

Nor were these the only adornments that would beautify L'Enfant's dream city. A Naval Column would be erected on the banks of the Potomac to celebrate the maritime glory of this yet-to-be-created force, and "five grand fountains" with a "constant spout of water" would be distributed throughout the capital. Each of the squares created by the angular intersections of the great boulevards would be named after a state—fifteen squares for the fifteen states then making up the union. "The center of each Square," L'Enfant explained, "will admit of Statues, Columns, Obelisks, or any other ornament such as the different States may choose to erect." One of the most ambitious elements of the plan, and one of the few that was never attempted in any form, was a canal that would run from the Potomac along the northern edge of the Grand Avenue to the foot of Jenkin's Hill. There it would turn south and form a Grand Cascade in front of the Federal House before slipping off into the Eastern Branch.

The visual impact of L'Enfant's imagined capital was breathtaking. High above the city rose the grand classical dome of the Federal House, which, L'Enfant boasted—his monarchical roots clearly showing—"would rear with a majestic aspect over the Country all around."[26] The elegant Grand Avenue and its gardens, sloping down from this seat of Congress toward the river, would focus all eyes on this, the people's house. Coequal would be the president's house, which L'Enfant conceived not as the relatively modest mansion known today as the White House but as a true presidential palace, on the same scale as Congress's Federal

FIGURE 32: Aerial view of the National Mall from Capitol Hill, ca. 1990, showing the geometrical patterns of L'Enfant's design

House. It too was placed on high grounds, with its own park sloping toward the river, accentuating its place at the apex of the republic's executive power. Standing at the foot of Washington's horseback statue (site of today's Washington Monument), one could look directly eastward, up the Grand Avenue toward the house of Congress looming above, or north, through the president's park toward his palace. Together the two edifices dominated the city's skyline, competing yet complementing each other, visual rivals that nevertheless come together in harmony.

There were other centers of beauty and grandeur in L'Enfant's plan. If the seat of Congress and the presidential palace commanded the city as a whole, the fifteen star-shaped plazas dominated their immediate surroundings. Spread evenly throughout the city, they were connected by perfectly straight avenues and clear lines of sight from one to the next. Each was unique, dedicated to one of the states, its history and its culture. Yet together they formed an interconnected network of local centers that reached to every corner of the capital. The houses of Congress and the president belonged to the entire city, but each neighborhood also possessed its own center, which belonged to it alone.

Taken as a whole, L'Enfant's plan was nothing less than the Constitutional power structure of the United States set in stone, pavement, trees, and shrubs. The two grand edifices embodying the pillars of federal power towered over the capital just as the institutions they housed reigned supreme over all others in the republic. Their power spread out into the capital through the straight grand avenues and green parks that radiated from them in every direction. Yet just as the central authority of the president and Congress was tempered by the local powers of the states, so it was in L'Enfant's capital. In each particular location the local plaza loomed as large as the great federal palaces, and together they lay like an all-encompassing net over the city. The great federal seats of power could no more break through the leveling powers of the states than the states could escape the federal power radiating from Jenkin's Hill and the President's Palace.

Versailles on the Potomac

Versailles, as we have seen, was royal supremacy writ large. In Louis XIV's garden all roads and all lines of sight converge on the royal palace atop the hill, just as all power emanates from it to every corner of the garden, and through it to the kingdom itself. L'Enfant's design for the American capital was created to accommodate a more complex power structure and is consequently more intricate. Whereas Le Nôtre's sole focus was on glorifying the king in his palace, L'Enfant needed to accommodate multiple centers of power, each different in nature, which operate in a mixture of tension and harmony. His success in doing just that, imagining a capital that would be didactic and metaphorical and yet a lively, beautiful, living city, is surely one of the greatest feats of city planning ever.

Yet for all the differences between the political orders celebrated in Versailles and in Washington, the similarities between the two designs overwhelm them all. The palaces on the hill with the formal gardens sloping down from them are an unmistakable echo of Versailles. So are the broad linear avenues converging on the centers of power from opposite directions, usually in symmetrical groups of three, a pattern familiar from Louis XIV's town and gardens. The extensive use of water effects, the canal, the great cascade planned before the Federal House, the star-shaped plazas where the avenues meet, and the "Statues, Columns, Obelisks" within them are all hallmarks of the royal absolutist architecture of old-regime France.

And more than anything it is the geometry of L'Enfant's design that makes his city an American Versailles. It is the most obvious feature of the plan, and the first thing that stands out from even the most cursory glance. To anyone looking at the plan, the avenues connecting the great and small squares and plazas serve as the sides of myriad geometrical shapes—isosceles triangles, right triangles, rhombuses, and rectangles—all neatly fitted together to form the surface of the city. The end result is a precise mathematical landscape made up of fixed, immutable, geometrical shapes, a city as Euclid might imagine it.

At Versailles, as we have seen, the geometrical order of the gardens presented the political order of monarchy as necessary, inevitable, and irresistible. Walking the garden paths, one would not just learn the ideology of the absolutist state, but physically experience it as part of the unshakable order embodied in geometry itself. Inevitably, L'Enfant's city—designed to serve a republic rather than an absolutist monarchy—differed markedly from the plan of Versailles. Where Louis XIV's capital pointed to a single glorious seat of authority, the American capital featured two coequal seats of power, as well as a network of lesser ones spread throughout the city. Yet no less than Versailles, the future city of Washington was a geometrical construct, visible proof that in the American capital—just as at Versailles—the political order is at its core a geometrical order.

In geometry, the sum of the angles of a triangle is equal to two right angles, and the sum of the squares of the sides of a right triangle is equal to the square of the hypotenuse, not because Euclid said so, but because the laws of the universe compel it. Consequently, the geometrical jigsaw puzzle that makes up L'Enfant's city does more than reiterate the divisions of power laid out in the Constitution. It indicates that the republican political order is not an artificial construct, but part of an inescapable universal order. The separation of powers between the legislative and executive branches of government, and the balance between federal authority and the autonomy of the states, are as rational, necessary, and irrefutable as a Euclidean proof.

And yet, presenting the rule of Louis XIV as part of an immutable universal order is one thing—doing the same for the American political structure is quite another. The Sun King, after all, was an absolute monarch who claimed divine right for his rule and acknowledged no limit to his powers. The American political order, in contrast, is founded on the Constitution—a man-made document, the result of exhaustive deliberations and compromises. Not only that, but the Constitution, so far from being immutable, was changeable by design: indeed, the first ten amendments, known as the Bill of Rights, were ratified in 1791, the same year L'Enfant was drafting his plan for the capital. Why then should the decrees of this very human and rather malleable document be presented as timeless and unassailable geometrical truths?

The answer is that L'Enfant's plan was not a formal expression of what the Constitution "actually was," but a public statement of how a certain political faction wanted it to be perceived. And for them, L'Enfant's design of the capital served as a master stroke of political propaganda. During the Constitutional Convention of 1787 and the ratification battles that followed, the men who perceived a need for the Constitution and advocated for its passage were known as Federalists.

Led by James Madison, Alexander Hamilton, and, most crucially, George Washington, they warned that the states were pursuing their own narrow interests, bringing the union to the point of dissolution. To forestall such a catastrophe, the Federalists pushed to replace the loose Articles of Confederation with a binding and enforceable Constitution. The document that was ultimately adopted was not all that they had hoped for (it granted too much power to the states), but they quickly rallied around it as the best that could be accomplished. Fearing a return to the chaotic early days of the republic, they were determined to present the Constitution as a done deal, an irreversible turning point for the union and a second founding for the nation.

It is in this context that we should consider L'Enfant's plan for the federal capital. For the Frenchman himself, it likely seemed a natural extension of Versailles: if Le Nôtre's gardens proclaimed French royal absolutism to be part of the universal rational order, why should his design claim any less for American republicanism? But for his Federalist supporters, and primarily for Washington, L'Enfant's city served a more specific political purpose. In presenting the Constitutional order as founded on the irrefutable natural order, they were making it that much more difficult to change or overturn it. Indeed, it would be all but unthinkable to revive the confederate structure of the Articles, or to establish a parliamentary democracy, in a city whose very design proclaims the supremacy of federal institutions and the separation of powers of Congress and the president.

It is impossible to say to what extent Washington and his fellow Federalists truly believed in the city's universalist message, and to what extent L'Enfant's plan was, for them, just a brilliant stroke of political promotion. But even if it was a feat of public relations, this much is certain: its message that the U.S. Constitution is founded on unchanging universal principles succeeded beyond what anyone could possibly have imagined. In 1882, nearly a century after its ratification, the historian George Bancroft waxed lyrically on Washington's suggestion that "the finger of Providence" had been at work in the Constitutional Convention and ratification process. In the creation of the U.S. Constitution, Bancroft wrote, Washington "saw the movement of the divine power which gives unity to the universe, and order and connection to events."[27] And while most historians have since turned against such hagiography, that still is how the Constitution is widely viewed in the American public and political spheres. It is the bedrock of the nation, founded on universal rational principles. And for all intents and purposes, it is also unchangeable.

That L'Enfant was proud of his design for the capital of the "American Empire" there is no doubt.[28] He was, in fact, so wedded to his plan that he found even small alterations intolerable. When his fellow surveyor Andrew Ellicott, at Washington's request, published a map of the projected city that differed in seemingly minor ways from L'Enfant's own, he declared the plan "most unmercifully spoiled."[29] When the president's house was, in later years, scaled back from its palatial dimensions, he judged the result "hardly suitable for a gentleman's country house," not to mention for "the chief head of a souverain people."[30]

L'Enfant's complaints might seem petty in retrospect, but there was more to them than the wounded vanity of a designer who had fallen in love with his own creation. L'Enfant believed that his design, like the political system it stood for, was not arbitrary, but the expression of a deep rational order, as fixed and unalterable as geometry itself. And just as one cannot alter the Pythagorean theorem without destroying the entire Euclidean edifice, so it was with the federal capital. Any attempt to tamper with the precise geometrical structure of the city would render its entire layout incoherent, meaningless, and, consequently, ugly. And so, even when faced with imminent dismissal and a crippling blow to his reputation and professional prospects, L'Enfant refused to compromise. In the kingdom of geometry there are no shades of gray. There is only truth or error.

The Battle of the Geometries

If even Ellicott's minor revisions to his design raised L'Enfant's ire, little wonder that Jefferson's radically different conception ignited in him what can only be described as an uncontrolled fury. Jefferson, in fact, had made two sketches of the federal city as he saw it, each suitable for a different location.[31] The first was enclosed in a memorandum he sent to the president dated November 29, 1790, reporting on the progress made in implementing the bill passed by Congress the previous summer ordering the acquisition of land for the federal capital. At the time, difficulties in negotiating the purchase of the land near Georgetown had convinced Jefferson that the capital would likely be built farther to the south and east, at the point where the Anacostia (or Eastern Branch) flows into the Potomac River. There, Jefferson proposed to build a city in which "no street be narrower than 100. feet, with foot-ways of 15. Where a street is long & level, it might be 120. feet wide." As for the size of the capital, he estimated that "1500 Acres would be required in the whole, to wit, about 300 acres for public buildings, walks, &c, and 1200 Acres to be divided into quarter acre lots."[32]

By American standards Jefferson's plan was unquestionably grand. Philadelphia, the largest city in the union, covered less than a thousand acres at the time, and no city in the land sported streets and avenues on the scale that Jefferson proposed.[33] But only a few months later, Washington achieved a breakthrough in his negotiations with the landowners farther to the west, and it once again appeared that the capital would be built along the north bank of the Potomac just east of Georgetown. Jefferson quickly sketched a plan suitable for the new location and delivered it to Washington at the end of March 1791. This sketch shows an even larger city than the previous one, measuring about one and a quarter by two and a quarter miles and covering roughly two thousand acres. Since no detailed notes survived on this draft, we can only assume that in Jefferson's mind the width of the streets and avenues remained as they were in his previous plan.

Jefferson's two plans differ in location and size, and in the overall shape and outline of the proposed cities. But even a single glance at them reveals that in their core design philosophy the two are not just similar, but practically indistinguishable: whether on the banks of the Anacostia or the Potomac, Jefferson was proposing a city defined by a regular rectilinear grid. Long avenues would stretch lengthwise, from one end of the city to the other, while shorter streets, also traversing the entire city between opposite sides, would intersect them at right angles and regular intervals. The end result would be a city designed as a uniform plane, made up entirely of square or rectangular blocks, all precisely the same size and shape, and lined up next to one another. If L'Enfant's city is a dazzling jigsaw puzzle of geometrical figures, dramatically arranged to form a city landscape, Jefferson's city is a uniform checkerboard in which all blocks are not just equal but identical.

Jefferson made no secret of his preference for a square rectilinear pattern for the capital city. "I should propose these [lots and squares] to be at right angles as at Philadelphia," he wrote in the report that accompanied his first plan, asking rhetorically: "Will it not be best to lay out the long streets parallel with the creek, and the other crossing them at right angle, so as to leave no oblique lots, but a single row which shall be on the river?"[34] It would, of course, be best, according to Jefferson. "Oblique lots," of the kind that enliven L'Enfant's plan, are clearly something to be avoided.

Nor was Jefferson's love of square blocks rooted in ignorance of alternative, more variable designs. During his years as minister to France he came to know not only Paris, his primary residence, but also other great European cities such as Amsterdam, Frankfurt, Strasbourg, and Milan, and wherever he went he made sure to purchase a city plan. Yet none of these famous cities, with their grand boulevards and adorned star-shaped plazas, impressed the secretary much: "They are none of them," he wrote dismissively to Washington, "comparable to the old Babylon, revived in Philadelphia & exemplified."[35] It's hard to know what Jefferson knew, or thought he knew, about the design of ancient Babylon, but the reference to Philadelphia makes clear what he had in mind, for the City of Brotherly Love, from its founding in the 1680s, was designed as a strict, uniform, rectilinear grid.

Jefferson's plan, to the modern eye, seems inoffensive enough. It was a large and ambitious metropolis by the standards of its time, sensibly designed, and inspired by the de facto capital, Philadelphia. What was it about it, then, that so inflamed L'Enfant that he would risk career and reputation to denounce its creator as a man of small mind and no imagination? Partly, no doubt, it was a matter of scale: Jefferson's capital, though larger than any existing American city, covered but a third of the territory of the Frenchman's grandiose city. This was much too small for L'Enfant, who never tired of reminding Washington that the city's size and grandeur must be proportionate to the greatness of the new American empire. And yet if size was his main concern, L'Enfant could have said as much without denouncing the entire plan as "defective" and "injurious" to the entire undertaking. It is, after all, one of the advantages of a rectilinear plan that it can easily be expanded to cover any area without fundamentally changing the design.

To some extent it may have been the relative prominence of the seats of federal power in Jefferson's design that, when compared to his own, concerned L'Enfant. In his own plan the Federal House and President's Palace dominate the landscape, drawing every eye from every corner of the capital, and indeed from miles beyond it. But nothing like this was evident in the secretary's plan. Jefferson did, to be sure, reserve certain areas in his city for the main seats of the federal government. In the memorandum attached to his first plan he suggests that two blocks should be consolidated for "the President's house, offices, and gardens," and one block reserved for "the Capitol," the seat of Congress that L'Enfant refers to as the Federal House.[36] In his second plan, which shows more detail, the president's house and the Capitol take up three blocks each. But compared to the monumental edifices with their grand avenues and expansive formal gardens that L'Enfant had in mind, Jefferson's proposal seemed rather shabby.

Yet even this can hardly explain the Frenchman's wrath. The Capitol and the president's house were certainly large in Jefferson's plan, taking up a significant portion of the city. In his first plan he allocated one-fifth of the total fifteen hundred acres for public buildings, and in his second plan the two main federal edifices take up six of the capital's thirty-three blocks—again, about a fifth. Since L'Enfant was proposing a larger city, he could easily insist on enlarging the grounds set aside for the grand public buildings without dismissing the plan as a whole. From whence, then, the outrage?

The true reason why L'Enfant was so offended by Jefferson's plan was not a matter of size or grandeur, but something even more basic, and so fundamental as to be almost invisible: it is the plan's geometry.

Jefferson's design was, to be sure, just as geometrical as the Frenchman's own: both were constructed of perfectly straight lines intersecting one another at precisely measurable angles. In both the surface of the city was composed of polygons, nestled alongside one another, whose area and circumference could be exactly calculated by the laws of Euclidean geometry. And yet the geometry of the two plans could not be more different.

L'Enfant's plan is made up of straight grand boulevards that traverse the capital diagonally in all different directions; Jefferson's city is made up of equally straight boulevards, but they are either parallel to one another or at precise right angles. Whereas the distances between L'Enfant's avenues vary with distance and location, Jefferson's parallel avenues are all at regular and fixed distances from one another. And whereas the boulevards in L'Enfant's plan come together in grand star-shaped plazas, Jefferson's intersections are simple right-angled crosses.

The difference between the two schemes goes deeper than aesthetic preference. L'Enfant's city, like Versailles, is a rigorously Euclidean landscape in which everything is exactly as it must be. Try moving one of the great avenues in L'Enfant's elegant design a little to the north or south, or altering its angle ever so slightly, and it is immediately apparent that it is practically impossible. The great hub of Federal Hall, with its twelve grand boulevards converging on it from every direction, is held in place by an elaborate network of interlocking plazas and intersections. And precisely the same is true of the president's house and each and every state-named square or plaza. Any attempt to move a single boulevard will disrupt the entire interconnected network, altering angles at intersections throughout the city and effectively destroying the elegant star-shaped plazas. L'Enfant's plan, in other words, is unchangeable. It is a world in which everything is inflexible and unalterable, a world whose intrinsic order is as unchallengeable and irrefutable as the truths of Euclidean geometry.

This, furthermore, is precisely the effect that L'Enfant was working to produce: the American capital embodied the nation's republican constitution just as Versailles represented French absolutist monarchy. By making it a geometrical dreamscape that could not be changed or challenged, L'Enfant was making sure that the young republic's political order would be viewed and experienced in the same way. In the American capital, the delicate balance of power between Congress, the president, and the states is as much an expression of the deepest order of the universe as the laws of Euclidean geometry.

In Jefferson's city there was nothing of the sort. Whereas in L'Enfant's plan all roads lead to the political power centers, in Jefferson's plan they lead nowhere in particular. They simply traverse the city from end to end at regular intervals, carving it up into identical blocks, each the same size and shape as the others. In L'Enfant's design the boulevards create a hierarchy between different spaces, from the Federal House and President's Palace at the top to the local plazas and the residential streets at the bottom. Jefferson's grand avenues treat all spaces equally, as all rectangular blocks are equal in shape and size and any given point in the city is equivalent to any other. Jefferson's imagined city is as flat as it is uniform, a landscape devoid of the political hierarchies that L'Enfant worked so hard to create.

It wasn't so much that the layout of Jefferson's city speaks of a more egalitarian order than the Frenchman would have preferred. L'Enfant, after all, was a committed republican who had risked his life for the proposition "that all men are created equal." It was, rather, that Jefferson's plan promoted no particular order at all. The gridded city has no focal points, no star-shaped plazas, and no grand avenues and parks extending from them. It has no center and no hierarchy, just a uniform rectilinear landscape from end to end. It is, in essence, the geometry of an empty mathematical landscape in which anything can be created—or nothing at all. L'Enfant's design was the Constitution set in stone, a political order made real, physical, and as irrevocable as geometrical truth. But Jefferson did not believe any of that: in a letter to Madison he once insisted that no generation is bound by the laws of its predecessors, and should instead write its own laws from scratch. Consequently his design, far from presenting a permanent Constitutional order, was laid out as an empty slate—a "tabula rasa" on which anything could be inscribed, erased, and reinscribed again in a completely different pattern.[37]

Take for example Jefferson's plan for the halls of Congress and the president's house. He did, as he makes clear in his second plan, make proper accommodations for them: each would occupy a sizable lot, larger than any other and made up of several standard blocks combined. This was as one would expect for the key institutions of the federal government, and in terms of size alone the design was giving them their proper due. Yet there was nothing in the city's plan as a whole that would determine where the two edifices should be located or how large they should be. Jefferson placed the future Capitol to the east of the president's house, and allotted each of them three blocks. But he could just as easily have switched their locations and moved either or both a few blocks in any direction. He could also have decided to assign them two, four, or six blocks instead of three, and to provide one of the two with a larger lot than the other. He could even have decided to eliminate one of the two, or both—all without affecting the uniform grid-like design of the city one bit. Such freedom is unthinkable in L'Enfant's design, in which each edifice is held in place by rigid interlocking geometrical patterns.

Just like L'Enfant's plan, Jefferson's design was geometrical, but its geometry was of a very different kind. Unlike the Frenchman's diagonals, triangles, and polygons, the identical rectangles marked not a fixed order, but the mathematical coordinates of space itself. The rectilinear coordinates do not define what might fill them: they are merely markers of an absolute and completely vacant space, a staked-out territory that could be filled up by anything. If L'Enfant's city is a Euclidean utopia, Jefferson's city is a Cartesian dream, an abstract mathematical space marked by the rectilinear grid waiting to be filled. In L'Enfant's city everything is as it must be; in Jefferson's city, anything is possible.

Seen in this light, Jefferson's capital of mundane, indistinguishable squares was, to L'Enfant, not merely drab and unimaginative, it was the betrayal of an ideal. To one walking the streets of such a capital, the political order, the relationship between the centers of authority and the balance between them, must appear changeable, impermanent, and easily revisable. In L'Enfant's design the government of the United States is an expression of immutable cosmic laws; in Jefferson's, the constitutional order of the republic feels as arbitrary as a passing fashion or the whim of an urban architect.

To see just how appalling Jefferson's capital must have appeared to L'Enfant, consider the contrast between Washington, D.C., and New York's borough of Manhattan as we know them today. Washington, whose modern layout closely follows the Frenchman's design, is still an immutable Euclidean metropolis. One can no more alter the location of the Capitol or the White House than one can move the royal palace at Versailles to the bottom of the valley and place it next to the Grand Canal. It is simply unthinkable, and would immediately destroy the coherence, purpose, and meaning of this elegant geometrical city, turning it into pointless chaos. But things are different in Manhattan, where a uniform Jeffersonian grid covers most of the island. Manhattan too has its grand structures: the Empire State Building, One World Trade Center, the New York Public Library. They are as famous as the capital's great monuments, or nearly so, and are as integral to our notion of New York as Capitol Hill is to our idea of Washington.

But suppose the library and the old skyscraper traded places, moving the classical NYPL to Thirty-Fourth Street and the Empire State Building to Forty-Second Street; or else that the two skyscrapers, one in lower Manhattan, the other in midtown, switched locations. How much would that change the city? The honest answer is—hardly at all. One could add new edifices, demolish old ones, or move them around without altering our experience of the city one bit. It would still be the same loud, vibrant, rectilinear metropolis that so many love, and just as many hate.

Exactly the same would be true of Jefferson's rectilinear capital, had it been built. The key governmental structures could still be built on an impressive scale, but there was nothing in the design of the city that would define their place in the political order. To a visitor walking by the Capitol or the president's house, the power dynamic between the two would appear as arbitrary as that between the capitalist commercial powers that raised the skyscrapers of Manhattan. This, L'Enfant would not condone. His capital was not merely a seat of the U.S. government, but a statement of its core principles and its rational underpinnings. Like many of his contemporaries, he believed that the United States was not just one more country on the world stage, but the bearer of universal principles of reason and liberty. His design, which made the balance of power prescribed in the Constitution into a key part of the universal rational order, expressed this faith as powerfully as anything could.

If L'Enfant were forced to conform to the secretary's plan, all of this would be lost. In the uniform grid of Jefferson's city universal principles of order dissolved into nothing, replaced by all-too-human whims and fashions. And so, in the end, it was not L'Enfant's vanity that caused him to risk his livelihood by taking on the wily and powerful Jefferson, nor was it an inflexible adherence to the style of design he had learned in his youth in France. It was, rather, his ardent belief in the founding principles of his adopted country. It was, in a word, his patriotism.

The Geometry of Union

Despite the irrecoverable damage done to L'Enfant's career, his plan did, in the end, carry the day. The city that we know today, with its grand diagonal boulevards converging on star-shaped plazas, and the monumental National Mall spreading westward from the majestic dome of the Capitol, was for the most part forged in the mind of the French firebrand. He is, without a doubt, the true father of the American capital.[38] To say that this outcome is surprising understates the case: L'Enfant served in his official post for less than one year, and in the space of that time managed to alienate not only Jefferson, but also the commissioners in charge of the capital's construction, who were his immediate superiors, and ultimately even his friend and ally the president. His dismissal was greeted with near-audible relief by those engaged in building the capital, and the reputation he earned for being arrogant, inflexible, and insubordinate haunted him for the rest of his years.

That the plan survived despite the disgrace of its creator is due in large part to President Washington, who was closely involved in every stage of the design and construction of the city that would bear his name. Washington liked L'Enfant's plan from the beginning, and his support did not waver even after its designer was dismissed. To his dying day Washington insisted, against considerable opposition, that the American capital would be built as a grand geometrical city inspired by Versailles, just as L'Enfant had envisioned it.[39]

It is possible that personal vanity played a role in Washington's unyielding support for L'Enfant's plan. By seeing it through, Washington ensured that his name would forever be associated with one of the greatest and most beautiful cities in the world. L'Enfant's city, after all, was meant to rival the great capitals of Europe in her magnificent architecture and monuments and to surpass them in her elegant design. Yet vanity alone cannot explain the president's steadfast support for building the geometrical capital. For a man like Washington, who deeply identified himself with the new American nation, even an act of self-glorification served a higher purpose.

Having seen the young republic struggle through the first decades of independence, Washington was forever haunted by the specter of disunion. The fear that the new nation he and his contemporaries had fought so hard to forge would disintegrate into a medley of weak and squabbling states was never far from his mind. This would be a disaster not just for America, he believed, but for the world, as it would signal the failure of the only state founded on the principles of liberty. To prevent this Washington insisted upon a strong federal government with powers to override the local interests of the states. Knowing his own popularity throughout the states, and his status as a symbol of shared struggle and common nationhood, he was not above encouraging a cult of personality if that would help cement the union. If he judged that an act of apparent hubris on his part was required to keep the union from splintering, then so be it.[40]

And that, it seemed to Washington, was precisely the case with the new federal capital that would bear his name. No less so than the Constitution itself, L'Enfant's geometrical capital was to Washington an antidote to the specter of disunion. Tightly designed and inflexible, the plan was a single carefully balanced whole that could not be altered, loosened, or disrupted. The centers of power—Federal House and the president's mansion—were held in place and bound together by the universal laws of geometry, as was every boulevard and state-named plaza. The city was an indivisible unit, and no part, great or small, could be moved or fly off on its own. And since the city's design replicated the careful balance of powers elaborated in the Constitution, it was a living demonstration that these powers were bound together in an indissoluble union. On the streets of the capital, thoughts of disunion were well-nigh impossible: the unity of the states and the federal government was as irrevocable as the truths of geometry.

And so, thanks to Washington's unwavering support, L'Enfant's paradoxical plan for a republican capital inspired by a royal garden lives on. It lives on not as a museum frozen in time, like Versailles or Schönbrunn Palace, but as the beating heart of a great nation and the embodiment of its spirit. To experience the power of the geometrical landscape of Washington, D.C., we do not need to transport ourselves back in time and imagine ourselves in the role of courtiers or ambassadors to Louis XIV or Peter the Great. All we need is to wander the streets of the American capital, walk through its broad, straight avenues, pass through the precisely measured intersections, and explore the star-shaped plazas—all overseen by grand edifices, monuments, and the glorious dome of the Capitol. It is all there. The constitutional republic is alive in Washington, D.C., as immovable, irresistible, and necessary as geometry itself.

Louis XIV is long gone, and his dream of royal supremacy has been tossed aside by the currents of history. Le Nôtre too is long dead, and his creations are preserved as mere shadows of their former selves, museum pieces frozen in time. Yet in the streets of Washington, D.C., the legacy of Versailles—its beauty, its power, and most of all, its geometry—lives on. Beneath the grand National Mall, the imposing Capitol Building, and the dignified White House, beneath the hum of a great administrative center, buzzing with activity, we can still detect the ghostly outlines of the Sun King's capital and its magical gardens. And if we stop and listen closely, we may yet hear the whisper of an obscure French monarch named Charles VIII, looking over a country villa outside Naples called Poggio Reale.

CONCLUSION: THE NON-EUCLIDEAN WORLD

The year was 1823, and János Bolyai (1802–1860), a young lieutenant in the service of the Habsburg Empire, was afire. A handsome and dashing Hungarian nobleman, Bolyai was known to his peers as a gifted swordsman, a habitual dueler, and the best dancer in the Austrian army. On one occasion, the story goes, he fought and defeated thirteen challengers in succession, playing the violin between rounds to steady his nerves. But it was not affairs of honor or the love of women that stirred his passions that fall: it was mathematics. For Bolyai believed he had made a mathematical discovery so momentous that nothing would ever be the same again. "All I can say," he wrote breathlessly to his father, "is that I have created a new and different world out of nothing." Remarkably, Bolyai was not wrong.[1]

At first glance, the details of Bolyai's work hardly seem to justify his audacious claims. Like many mathematicians before him, he was investigating the status of Postulate 5 of the *Elements*, Euclid's masterpiece that founded the science of geometry back in 300 B.C.E. "If a straight line falling on two straight lines makes the interior angles on the same side less than two right angles," it stated, "the two straight lines, if produced indefinitely, meet on that side on which are the angles less than the two right angles." This implied that parallels—lines that never meet—are those whose interior angles add up to precisely two right angles.

No one doubted that Postulate 5 was important. It made it possible to move angles around a plane by drawing parallels, a critical tool in many of the *Elements*' key propositions.[2] It also *seemed* true, and for over two millennia no mathematician had ever doubted that it was. But unlike Euclid's other postulates and common notions, it was not exactly self-evident. Other postulates posited things such as "all right angles are equal to each other" and that one can always "draw a straight line from any point to any point." The fifth, however, was far more complex, if not convoluted, looking more like a theorem that must be proved than a postulate that should be assumed. The young Hungarian was only the latest in a long line of commentators dating back to the ancient world to try to cleanse this unseemly blemish on the pristine beauty of Euclidean geometry.

But Bolyai's approach was different from that of his predecessors: instead of finding a way to reconcile the problematic postulate with the rest of geometry, he chose to abandon the postulate altogether and proceed without it. Starting with all of Euclid's assumptions except the troublesome fifth, he proceeded step by deductive step until he created an entirely new geometry, one that differed profoundly from the only one considered possible—or even conceivable—until that day. In Euclid's world the sum of angles in a triangle would always add up to 180 degrees, but in Bolyai's world they never quite reached that. In fact, the sum of the angles wasn't even fixed, and as the area of a triangle increased, the sum of its angles shrank. In Euclid's world figures could be scaled, meaning that the precise same figure could be larger or

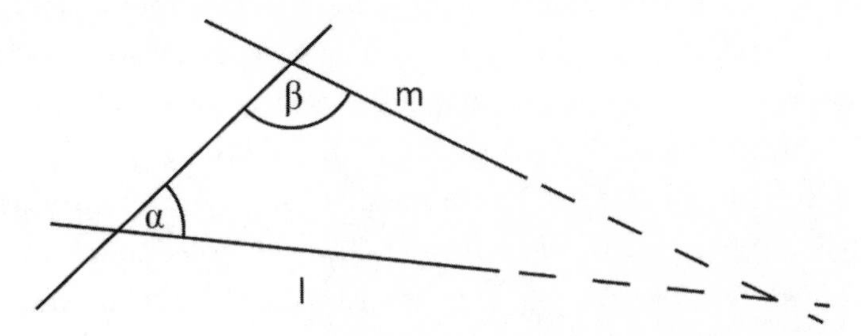

FIGURE 33: **Euclid's Postulate 5: if the sum of the interior angles α and β is less than two right angles, then the lines *l* and *m* will meet on that side**

smaller. But not so in Bolyai's: the angles in a figure changed as it was reduced or enlarged, and scaling was therefore impossible.[3]

Finally, Bolyai discovered that the trigonometric equations that describe non-Euclidean space are not fully determined, but contain an arbitrary constant he called "*k*." Choosing a different value for *k* leads to a different set of equations, or, in other words, a different geometry. Bolyai himself did not draw the conclusion that he had discovered a multiplicity of geometrical worlds, each corresponding to a different *k*, but others soon did. In the 1840s Carl Friedrich Gauss (1777–1855) discovered that each of Bolyai's geometries applied to a different curvature of space, and a decade later his student Bernhard Riemann (1826–1866) expanded new geometries to higher dimensions of any order. The verdict was clear: There was not just one non-Euclidean geometry, but an infinity of them, each different from the others and incompatible with them. And they were all equally true.[4]

To those who believed that the world was founded on a unique and necessary geometrical order, it was a crushing blow. Ever since Euclid, geometry had been viewed as a single system of inescapable truths, unchallengeable and indisputable, forced on the mind by rationality itself. Ever since Brunelleschi and Alberti, this single order was seen to shape physical reality, from the deep structure of matter to the motions of the heavens. And from Charles VIII to Louis XIV, and on to American republicans and British imperialists, those intent upon establishing immutable regimes had sought to anchor political order in this geometrical order.

But Bolyai's discovery threw these foundations into chaos. If the young Hungarian was right, then Euclidean geometry was no longer the single necessary and true order, but just one geometry of innumerable possible ones, each incompatible with the others. The Euclidean world, it turned out, despite its impressive pedigree and lofty pretensions, has no more claim to being the one true order than any other world forged by the strange geometries described by Bolyai. And if this is the case, then the prodigious efforts expended on founding social and political institutions on the eternal truths of geometry come to seem pointless. Yes, the Euclidean patterns of Versailles do uphold the rule of kings, and those of Washington speak eloquently of a republican balance of powers. But if Euclid's is just one of many possible geometrical orders, each equally true, then one might begin to suspect that alternative geometries may suggest very different—and perhaps opposite—lessons.

It was a powerful, perhaps even devastating blow to the old and beautiful dream of a geometrical universe. But it was far from the first. For, in fact, the geometrical world had been subjected to one attack after another almost from the moment of its inception. Some of its most persistent critics were those who doubted the very premise that the world was mathematically ordered. Among them were traditional Aristotelians, who held firm to the medieval tradition that viewed mathematics as inadequate to describe the diversity and complexity of the world. Accordingly, when Galileo was condemned by the Church in 1633 it was not solely for the crime of contradicting scripture by setting Earth in motion; it was also for the audacity of declaring that mathematics was the key to true knowledge of the world.

And it was not only conservative churchmen, but also forward-looking reformers who contested Galileo's dictum that the world is written in the language of geometry. Francis Bacon (1561–1626), Lord Chancellor of England and prophet of the new science, warned of the dangers of overreliance on mathematics. The true path to knowledge, he insisted, lay in the systematic accumulation of laborious empirical observations, not in elegant mathematical formulations. When some decades later the Royal Society of London adopted his outlook, it established a strong empirical strain that characterized English science into the nineteenth century. Outside of scientific circles, adherents of Romanticism denounced the cold, impersonal mechanisms implicit in the geometrical vision of the world. To counter the Euclidean geometries of Versailles they planted naturalistic English gardens on gentlemanly estates and in urban parks. They brought wilderness to the heart of cities and established natural preserves to highlight untamed nature's sublime creations.

Other critics shared the belief that the universe was deeply mathematical, but suggested that its deep order was not fully captured by Euclidean geometry. Starting in the early seventeenth century, mathematicians began exploring infinitesimal methods, based not on strict reasoning from self-evident postulates, but on a keen intuition of the atomic structure of nature. The new methods were powerful and effective in seeking and establishing new results, but they were also paradoxical: their core assumption, that the continuum is composed of indivisible parts (e.g., a line of points, a surface of lines) ran up against the well-established paradoxes of Zeno and the problem of incommensurability, both known since antiquity. These contradictions ruled out the new methods for traditional mathematicians, who believed the field should be perfectly rational and self-consistent, but the infinitesimalists were undeterred. In their eyes the clear advantages of the new approach more than counterbalanced any deficiencies it might have in terms of Euclidean purity. Consequently the infinitesimalists' world was fully mathematical, to be sure, but also more flexible, less deterministic, and less hierarchical than the one proposed by the champions of Euclid.[5]

Then, in the 1780s, the government of the young United States began experimenting with geometrical ordering of the vast expanses of North America, which it viewed as its rightful inheritance. Instead of relying on Euclidean patterns, however, the Americans turned to Cartesian geometry, with its rectilinear coordinates and regular grid pattern. Whereas Euclidean geometry defines a fixed, irrevocable order, the Cartesian variety defines only an empty mathematical space, open to any possibility. It was the proper geometry for the promise of the New World.[6]

Each of these critiques proposed different orderings of the world, and all of them challenged the Euclidean geometrical vision. Yet none was more devastating than the discovery of non-Euclidean geometry. This is because unlike the other critics, Bolyai did not challenge the traditional assumptions of geometry. On the contrary, he followed them religiously. Like Euclid, he believed that geometry must be founded on self-evident assumptions about points and lines, and that all results must be deduced from them, step by logical step. Like Euclid, he too proved irrefutable theorems and used them to prove still others. And again like Euclid, he too thereby created a geometrical world composed of interlocking truths that were interdependent, hierarchically layered, and entirely incontestable. Bolyai, it might be said, was as Euclidean as Euclid himself.

Except that, when all was said and done, Bolyai had reached the precisely opposite conclusion from Euclid and his admirers. Instead of establishing a single, necessary, indisputable order that governs everything eternally, he arrived at an infinity of possible orders, each as true and rational as the others. Each of these might accord with our world or it might not, might describe our reality or an alternate one, or might exist side by side with other, conflicting geometrical orders. It follows that any attempt to ground a political order—absolutist, republican, or any other—on Euclidean principles is entirely pointless. The world, for all we know, may be ordered very differently. Previously it had been geometry's critics who challenged its ability to correctly describe the world; but with Bolyai, it was geometry itself that proclaimed that geometrical truths might not apply to the world, and that incompatible truths could and did exist side by side. Nothing could be more devastating for the dream of a single incontestable rational world.

Looking at our present-day world through the eyes of geometry, it is easy to detect the imprint of Bolyai's challenge to the geometrical universe. Few indeed are those today who yet believe that human societies can or should be governed by eternal rational principles that apply to all alike. It has become commonplace to argue that different cultures adhere to their own cultural truths, as do different religions, different nationalities, ethnicities, genders, and sexual orientations. Indeed, the notion itself of a single objective truth has come under attack in the politics of twenty-first-century America, as the conservative tribe of Fox News watchers vies with the liberal tribe of MSNBC and CNN adherents. Each lives in its own bubble of seemingly self-consistent truths, which are entirely incompatible with those of the opposition. It is a fragmented world that we inhabit, composed of many different and contradictory truth-worlds that exist side by side. Each has its own assumptions, its own reasoning, its own devoted followers, and—seemingly—its own facts. It is a world that has lost faith in universal rational principles, settling instead for a multiplicity of incompatible truths. It is, in other words, a world in which Euclidean certainties have been shattered and the multiple truths of Bolyai's non-Euclidean geometry are made manifest in our social and political lives.

We have, without a doubt, come a long way from the geometrical determinism of Versailles, where Euclidean principles enforced immutable social hierarchy and royal supremacy. And yet the legacy of Louis XIV's kingdom within a kingdom is still with us. This is the case not only in our gardens and city streets but also in our governments, state bureaucracies, and civil service. The orderly hierarchy of local, regional, and national government, for example, recapitulates the "natural" hierarchy of power on display at Versailles; the smooth flow of directives from decision makers at the top tiers of state bureaucracies to enforcers at the bottom is a real-life manifestation of Versailles's ideal of unhindered power descending from the palace on the hill to every corner of the gardens.

We are almost as far removed from the Enlightenment optimism of L'Enfant's federal city, where ubiquitous geometrical patterns guaranteed timeless republican principles. And yet the careful balance between the different branches and levels of government, so elegantly written into the city's streets and plazas, remains the core of the American governmental system. More broadly, the universal ideals Americans hold dear, such as equality and inalienable human rights, also have their roots in L'Enfant's geometrical vision of a harmonious world governed by ubiquitous and eternal principles. And so, although the grand claims of the geometrical age appear hopelessly out of touch with our multicultural world, and though their belief that the truth is one, rational, and eternal teeters under a constant barrage of criticisms, it seems we simply cannot do without these principles. If relativists from left and right attack the old mainstream "center," it is those claims to universal values and timeless rationality they are attacking. But in the end, even the fiercest critics of universal reason cannot do without the claims encoded in the Euclidean geometries of Le Nôtre and L'Enfant.

Meanwhile, even as the Euclidean vision of the world seems perpetually under attack, on the most famous avenue in the world its glory has hardly dimmed. Paris's Champs-Élysées, originally laid out by Le Nôtre as a westward extension of the Tuileries gardens, has become in the space of two centuries the sparkling heart of the City of Lights. Broad and arrow-straight, it is bracketed by a massive triumphal arch (the Arc de Triomphe) in the west and a grand plaza with an Egyptian obelisk (the Place de la Concorde) at its center in the east, both perfectly aligned with the avenue's central axis. A smaller triumphal arch (the Carrousel) continues the same line past the Tuileries gardens to the east, while a symmetrical twelve-pointed plaza known as the Étoile (the Star) surrounds the Arc de Triomphe, marking the avenue's western end. At the height of its brilliance in the nineteenth century the Avenue des Champs-Élysées was the very embodiment of a great ceremonial axis—the glorious ideal to which grand avenues from Unter den Linden in Berlin to Kings Way in New Delhi could aspire but never quite reach. Even then, however, the avenue was not done evolving and expanding, its fortunes intertwined with the nation's history.

In 1970 the Étoile became the Place Charles de Gaulle, named after the hero of Free France and the nation's greatest postwar president. The western part of the Avenue de la Grande-Armée—extending the Champs-Élysées beyond the Étoile—became the Avenue Charles de Gaulle, and the axis was then extended even farther, beyond the Seine to La Défense, where it was capped with one more perfectly aligned arch. Officially named the Grande Arche de la Fraternité, it was appropriately inaugurated in 1989, on the bicentennial of the French Revolution, by François Mitterand—socialist, former resistance leader, and then president of the republic.

These latest extensions created what is known today as L'Axe Historique (the Historical Axis), making the Champs-Élysées the midsection of a perfectly straight line extending from the Louvre to La Défense. And a fitting name it is, for the "Axe" is but the latest chapter in a history that has lasted more than three centuries: the monarchy, the Revolution, two Napoleons, De Gaulle, and Mitterand had all left their mark on the grandest avenue in the world. The Champs-Élysées had become a geometrical manifestation of the power and glory of the French state, through all its evolutions and revolutions.

And so, even as President Mitterand was installing the arch at the western end of the Axe Historique, the architect I. M. Pei, working in the Louvre courtyard, made one last addition to its eastern origin: An equestrian statue of Louis XIV lined up with the arches, the obelisk, and the famous avenue. It was a fitting tribute to the king whose geometrical vision launched the avenue and is still imprinted on the gardens he loved and the city he feared. In Paris, as in Washington, the geometrical dream of Versailles lives on.

NOTES

INTRODUCTION

1. Accounts of Louis XIV's visit to Vaux-le-Vicomte can be found in Jean-Christian Petitfils, "Vaux-le-Vicomte. Le château de jalousie," in *Versailles, Le pouvoir de la pierre*, ed. Joël Cornette (Paris: Éditions Tallandier, 2006), 61–69; Paul Morand, *Fouquet, ou Le Soleil offusqué* (Paris: Gallimard, 1961), 87–103; and Charles Drazin, *The Man Who Outshone the Sun King: The Rise and Fall of Nicolas Foucquet* (London: William Heinemann, 2008), 217–24.
2. On the accomplishments of Hippocrates of Chios and his contemporaries see Ronald Calinger, *A Contextual History of Mathematics* (Upper Saddle River, NJ: Prentice Hall, 1999), 85–93.
3. Ancient tradition credited Euclid with a number of lost works in addition to the *Elements*, so it is of course possible that he did come up with new mathematical results that did not survive. But the basis of his fame and towering reputation is the *Elements*, a work that systematizes known results but adds no new ones.
4. The title *Elements* did not indicate that the work was what we call "elementary," or intended for beginners. It simply meant that it is a systematic presentation of a

field, founded on first principles. Euclid may not have been the first to author an "Elements" of geometry, as both Hippocrates of Chios and Eudoxus of Cnidus are credited with authoring their own "Elements." Their works may have been among Euclid's sources, but they did not survive and we do not know what they contained. It is Euclid's *Elements* that, already in antiquity, became the benchmark of logical exposition and the foundation of geometry's unmatched intellectual authority.

5. Hippocrates of Chios (ca. 460–380 B.C.E.), for example, used them in calculating the area of lunes, and Eudoxus of Cnidus (ca. 390–337 B.C.E.) applied them in comparing volumes of spheres and pyramids.
6. The exception to the rule of simple, self-evident postulates is the infamous Postulate 5, which reads: "That if a straight line falling on two straight lines make the interior angles on the same side less than two right angles, the two straight lines, if produced indefinitely, meet on that side on which are the angles less than the two right angles." Dissatisfaction with this strange postulate, and attempts to "prove" it based on simpler ones, led in the nineteenth century to the discovery of non-Euclidean geometry.
7. All quotes from the *Elements* are from *Euclid's Elements: All Thirteen Books in One Volume*, ed. Dana Densmore, trans. Thomas L. Heath (Santa Fe, NM: Green Lion Press, 2002).
8. Q.E.D. stands for *quod erat demonstrandum*, meaning "that which was to be proven." It marks the end of a proof, and was used (in its Greek form) by Euclid.
9. Geometers such as Euclid and Eratosthenes of Cyrene thrived in Ptolemaic Egypt, Archimedes (ca. 287–212 B.C.E.) was supported by King Hiero II of Syracuse, and Apollonius of Perga (ca. 200 B.C.E.) was apparently patronized by the kings of Pergamon in Asia Minor. For a wonderful account of atomism, from its origins in the ancient world to its impact on the modern one, see Stephen Greenblatt, *The Swerve: How the World Became Modern* (New York: W. W. Norton, 2012).
10. For more on Alberti, see chapter 2.
11. On Copernicus and his geometrical vision of the heavens, see Thomas S. Kuhn, *The Copernican Revolution: Planetary Astronomy in the Development of Western Thought* (Cambridge, MA: Harvard University Press, 1957).
12. Galileo Galilei, "Excerpts from *The Assayer*," in *Discoveries and Opinions of Galileo,* ed. and trans. Stillman Drake (New York: Anchor Books, 1957), 237–38. The original Italian was first published in Rome in 1623 as *Il Saggiatore*.
13. Bernard Le Bovier de Fontenelle, *Histoire du renouvellement de l'Académie royale des sciences en mdcxcix et les éloges historiques* (Amsterdam: Pierre du Coup, 1719–20), vol. 1, 14, quoted in J. L. Heilbron, "Introductory Essay," in *The Quantifying Spirit in the Eighteenth Century*, ed. Tore Frängsmyr, J. L. Heilbron, and Robin E. Rider (Berkeley: University of California Press, 1990), 1–25, p. 1.

1: THE MIRROR IMAGE

1. On the origins of the baptistery see Ferdinand Schevill, *Medieval and Renaissance Florence*, rev. ed., vol. 1 (New York: Frederick Ungar, 1961; repr. Harper Torchbooks / Academy Library, 1963), 241–42. On the belief that Florence was founded by

Julius Caesar and his veterans see Hans Baron, *The Crisis of the Early Italian Renaissance* (Princeton, NJ: Princeton University Press, 1955; rev. ed. 1966), 49–52.

2. The dating of Brunelleschi's experiment is based on the fact that in 1413 the humanist Domenico da Prato refers to him as "the perspective expert, ingenious man, Filippo di Ser Brunellesco, remarkable for skill and fame." See Ross King, *Brunelleschi's Dome: How a Renaissance Genius Reinvented Architecture* (New York: Bloomsbury, 2000), 34.
3. On the competition for the baptistery's doors see Richard Krautheimer and Trude Krautheimer-Hess, *Lorenzo Ghiberti* (Princeton, NJ: Princeton University Press, 1956), chap. 3, as well as King, *Brunellechi's Dome*, chap. 2.
4. The quote is from Mariano di Jacopo Taccola, "From a Record of a Speech by Brunelleschi (1420s)," in *Brunelleschi in Perspective*, ed. Isabelle Hyman (Englewood Cliffs, NJ: Prentice-Hall, 1974), 31.
5. On the difference between Brunelleschi and Ghiberti in their approach to the project see Antonio di Tuccio Manetti, *Life of Brunelleschi*, trans. Catherine Engass, ed. Howard Saalman (University Park: Pennsylvania State University Press, 1970), 46–48.
6. Today, Ghiberti's bronze panels are acknowledged as one of the great masterpieces of the Renaissance, adorning the eastern doors of the baptistery as part of what Michelangelo called "the gates of paradise." Brunelleschi's design can be seen as well, preserved in Florence's Bargello museum.
7. Among the most notable of the Florentine humanists of Brunelleschi's generation were the classicist Niccolo Niccoli (1364–1437), the historian Leonardo Bruni (1370–1444), and Poggio Bracciolini (1380–1459), a papal secretary and polymath who rediscovered Lucretius's *De Rerum Natura* in a remote monastic library.
8. For a lively and insightful account of early Italian humanism, and of Poggio Bracciolini in particular, see Stephen Greenblatt, *The Swerve: How the World Became Modern* (New York: W. W. Norton, 2012).
9. On Brunelleschi's sojourn in Rome, see King, *Brunelleschi's Dome*, chap. 2.
10. See Giorgio Vasari, "Filippo Brunelleschi," in Vasari, *The Lives of the Artists*, trans. and ed. Julia Conaway Bondanella and Peter Bondanella (Oxford: Oxford University Press, 1991), 118.
11. Quotes are from Manetti, *Life of Brunelleschi*, 52–54. On the surveying methods described by Fibonacci and likely used by Brunelleschi see Martin Kemp, "Science, Non-Science and Nonsense: The Interpretation of Brunelleschi's Perspective," *Art History* 1, no. 2 (1978): 134–61, pp. 143–46; and Jehane R. Kuhn, "Measured Appearances: Documentation and Design in Early Perspective Drawings," *Journal of the Warburg and Courtauld Institutes* 53 (1990): 114–32, pp. 117–22.
12. Manetti, *Life of Brunelleschi*, 52.
13. Manetti, 50.
14. The painting has since been lost, but Manetti provides a detailed description in his *Life of Brunelleschi*, 43–44.
15. The diameter of a Venetian ducat was around two centimeters, or four-fifths of an inch. The circumference, accordingly, was slightly more than six centimeters.

16. On Brunelleschi's experiment see Samuel Y. Edgerton, Jr., *The Renaissance Rediscovery of Linear Perspective* (New York: Basic, 1975), 124–29; John White, *The Birth and Rebirth of Pictorial Space* (London: Faber and Faber, 1967), 113–21; and David Wootton, *The Invention of Science: A New History of the Scientific Revolution* (New York: HarperCollins, 2015), 165–72. Manetti's account can be found in his *Life of Brunelleschi*, 44.
17. The word is derived from *perspectiva ars*, the Latin term for the science of optics. Some scholars argue that the principles of perspective had been known in antiquity, and some surviving paintings do indeed show an awareness of the importance of adjusting the size of an object to its intended distance from the viewer. For example, Vitruvius, the ancient authority on architechure, explains in *De Architectura* the principles of stage-set paintings known as *scenographia*, which some argue resemble linear perspective. There are, however, no surviving works from the ancient world in which perspective is applied systematically to an entire composition.
18. See King, *Brunelleschi's Dome*, 34.
19. Quoted in Michael Baxandall, *Painting and Experience in Fifteenth Century Italy: A Primer in the Social History of Pictorial Style* (New York: Oxford University Press, 1972; 2nd ed., 1988), 124–25.
20. On the painting's mention in the inventory of Lorenzo the Magnificent's effects after his death in 1492 see White, *Birth and Rebirth of Pictorial Space*, 119. The importance of mirrors to Brunelleschi's experiment was first noted by his younger contemporary the Florentine architect and sculptor Filarete. See Wootton, *Invention of Science*, 165.
21. For more about the significance of mirrors in Brunelleschi's experiment see Wootton, 169–72.
22. For detailed discussions of Brunelleschi's experiment see Edgerton, *Renaissance Rediscovery of Linear Perspective*, 143–52; White, 113–17; and Wootton, 165–72.
23. On Brunelleschi's design of the Church of San Lorenzo, and his adherence to strict geometrical proportions that he believed were derived from the ancients, see H. W. Janson and Anthony F. Janson, *History of Art: The Western Tradition*, 6th ed., vol. 2 (Upper Saddle River, NJ: Pearson / Prentice-Hall, 2004), 421–23.
24. For a reconstruction of Brunelleschi's painting of the Piazza della Signoria see White, *Birth and Rebirth of Pictorial Space*, 117–21. Wootton also discusses it in *The Invention of Science*, 166. Manetti's description can be found in his *Life of Brunelleschi*, 44–46.
25. On Brunelleschi's and Donatello's influence on Masaccio, and his journey to Rome, see Vasari, "Life of Masaccio," in *Lives of the Artists*, 101–109. The quote is from 109.
26. It hardly needs saying that Brunelleschi's lost painting of the baptistery, which he used in his 1413 experiment, was also constructed in accordance with the principles of linear perspective. Whether that painting should be considered a work of art or merely an experimental device is debatable.
27. The discussion of the two paintings follows Edgerton, *Renaissance Rediscovery of*

Linear Perspective, 27. Edgerton calls the practice of presenting all heads at a fixed level "isocephaly."

2: THE GEOMETRICAL CODE

1. On the Counts Alberti and their taming at the hands of the Florentine Commune see Ferdinand Schevill, *Medieval and Renaissance Florence*, vol. 1 (New York: Harper Torchbooks / Academy Library, 1963), chaps. 6, 7.
2. On the clash between the Alberti and the Albizzi see Schevill, *Medieval and Renaissance Florence*, vol. 2 (New York: Harper Torchbooks, 1963), 339–41.
3. Anthony Grafton, *Leon Battista Alberti: Master Builder of the Italian Renaissance* (New York: Hill and Wang, 2000; repr. Cambridge, MA: Harvard University Press, 2002).
4. Alberti arrived as a member of the entourage of Pope Eugenius IV, who was sheltering in Florence from an insurrection by his enemies in Rome.
5. Alberti wrote *Della Famiglia* between 1433 and 1439.
6. Leon Battista Alberti, *On Painting*, trans. Cecil Grayson (London: Penguin, 1999), 34.
7. Alberti, *On Painting*, 37.
8. Alberti, *On Painting*, 48.
9. Alberti, *On Painting*, 53.
10. Alberti, *On Painting*, 54.
11. The following discussion of Alberti's construction and use of the distance point is based on the interpretation in Edgerton, 44–49.
12. Alberti, *On Painting*, 59.
13. Alberti calls this procedure "circumscription"—"the process of delineating the external outlines of a painting" and defining its space before painting in the actual figures and objects. See Alberti, *On Painting*, 65.
14. Alberti, *On Painting*, 35.
15. Alberti, *On Painting*, 65.
16. The impact of *Della Pittura* extended to practically all Renaissance painters. For examples of works shaped by Alberti's treatise, see, for instance, the Sienese painter Neroccio de'Landi's *Annunciation*, dating from around 1475 and discussed in Edgerton, 55, and Filippo Lippi's *Madonna with the Child and Scenes from the Life of St. Anne*, painted in 1452, discussed in John White, *The Birth and Rebirth of Pictorial Space* (London: Faber and Faber, 1967), 182–83.
17. On Alberti's participation in the intellectual life of Lorenzo de Medici's circle see Grafton, *Leon Battista Alberti*, 337. On Alberti guiding Lorenzo and his friends in Rome in 1471 see Grafton, 258, and Martin Kemp, introduction to Alberti, *On Painting*, 1.
18. Quoted in Kemp, introduction to *On Painting*, 1.
19. For the full text and English translation of *Descriptio Urbis Romae* see Leon Battista Alberti, *Delineation of the City of Rome*, eds. Mario Carpo and Francesco Furlan (Tempe: Arizona Center for Medieval and Renaissance Studies, 2007). See also Robert Tavernor, foreword to Kim Williams, Lionel March, and Stephen

R. Wassell, eds., *The Mathematical Works of Leon Battista Alberti* (Basel: Birkhäuser, 2010), ix.

20. On Alberti's involvement in Nicholas V's rebuilding of Rome, see Grafton, 295–315, and Kemp, introduction to *On Painting*, 7.
21. The quote is from the earliest extant English translation of *De Re Aedificatoria*, by James Leoni in 1726. See Leon Battista Alberti, *Ten Books on Architecture [De Re Aedificatoria]*, ed. Joseph Rykwert, trans. Cosimo Bartoli (Italian) and James Leoni (English), book IX (London: Alec Tiranti, 1955) 194.
22. Alberti, *Aedificatoria*, book VII, 138.
23. The discussion is in Alberti, *Aedificatoria*, book VII, 138–39.
24. Alberti, *Aedificatoria*, book IX, 190–91.
25. Alberti, *Aedificatoria*, book VII, 139.
26. On the three classical orders being an "imitation of nature" see Alberti, *Aedificatoria*, book IX, 195–96, 200–201.
27. Alberti, *Aedificatoria*, book IX, 207.
28. The first to point out the contrast between the map at the Loggia and the Map with Chain was the great Renaissance historian Hans Baron, in *The Crisis of the Early Italian Renaissance* (Princeton: Princeton University Press, 1955; rev. ed. 1966), 171–72, followed by Edgerton in *The Renaissance Rediscovery of Linear Perspective*, 7–10. On the Map with Chain see also L. D. Ettinger, "A Fifteenth-Century View of Florence," *The Burlington Magazine* 94, no. 591 (1952): 160–67; and P.D.A. Harvey, "Local and Regional Cartography in Medieval Europe," in *The History of Cartography*, vol. 1, ed. J. B. Harley and David Woodward (Chicago: University of Chicago Press, 1987), 464–501, esp. 465, 477.

3: ROYAL GEOMETRIES

1. Charles VIII to the cardinal of Bourbon, March 28, 1495. Quoted in Kenneth Woodbridge, *Princely Gardens: The Origins and Development of the French Formal Style* (New York: Rizzoli International Publications, 1986), 40.
2. Quoted in Woodbridge, *Princely Gardens*, 39, and in Robert J. Knecht, *The French Renaissance Court, 1483–1589* (New Haven, CT: Yale University Press, 2008), 2.
3. Estimates of the size of Charles's army vary, ranging from 18,000 by Geoffrey Parker in *The Military Revolution: Military Innovation and the Rise of the West, 1500–1800* (Cambridge, UK: Cambridge University Press, 1988), 9, to the highly improbable 60,000 cited in Ferdinand Schevill, *Medieval and Renaissance Florence*, vol. 2 (New York: Harper Torchbooks, 1963), 436. The intermediate number of 27,000 is taken from Max Boot, *War Made New: Technology, Warfare, and the Course of History, 1500 to Today* (New York: Gotham Books, 2006), 2.
4. On Charles VIII's army and artillery, and his invasion of Italy, see Boot, *War Made New*, 1–6; Schevill, *Medieval and Renaissance Florence*, vol. 2, 435–38; and Parker, *Military Revolution*, 9–10.
5. This quote from Macchiavelli's *Florentine History* is from Felix Gilbert, "Machiavelli," in *Makers of Modern Strategy: From Macchiavelli to the Nuclear Age*, ed. Peter Paret (Princeton: Princeton University Press, 1986), 21.

6. Quoted in Woodbridge, *Princely Gardens*, 40.
7. A reproduction of the engraving can be found in Woodbridge, 36.
8. On monastic gardens see Franklin Hamilton Hazlehurst, *Jacques Boyceau and the French Formal Garden* (Athens: University of Georgia Press, 1966), 9.
9. The *Très Riches Heures du Duc de Berry* was created for Jean, Duke of Berry, by the Limbourg brothers between 1412 and 1416 and completed by other artists later in the century. It is considered an outstanding example of late Gothic manuscript illumination.
10. Hazlehurst, *Boyceau*, 9–10. The image from the Duke of Berry's Book of Hours can be found in figure 4.
11. On de' Crescenzi, Charles V, and early royal gardens see Hazlehurst, *Boyceau*, 10–11, and Woodbridge, *Princely Gardens*, 13–17. On the gardens of the Hôtel Saint-Pol see Woodbridge, 16–17.
12. On the Florentine architects of Poggio Reale see Woodbridge, 35.
13. On the reading of *De Re Aedificatoria* to Lorenzo's circle in 1486 see Anthony Grafton, *Leon Battista Alberti: Master Builder of the Italian Renaissance* (New York: Hill and Wang, 2000; repr. Cambridge, MA: Harvard University Press, 2002), 286.
14. Leon Battista Alberti, *Ten Books on Architecture [De Re Aedificatoria]*, ed. Joseph Rykwert, trans. Cosimo Bartoli (Italian) and James Leoni (English), book IX (London: Alec Tiranti, 1955), 193.
15. On Italian Renaissance gardens as representations of the inherent geometrical order of the world see Luke Morgan, "Design," in *A Cultural History of Gardens in the Renaissance*, ed. Elizabeth Hyde (London: Bloomsbury Academic, 2013), 19.
16. For more on Lorenzo's interest in gardens see Michel Baridon, *A History of the Gardens of Versailles*, trans. Adrienne Mason (Philadelphia: University of Pennsylvania Press, 2008), 10.
17. Alberti, *Aedificatoria*, book IX, 193.
18. Quoted in Hazlehurst, *Boyceau*, 96–97, *n*12.
19. Alberti, *Aedificatoria*, book IX, 195.
20. For more on how gardens, and in particular geometrical gardens, promote the creation of order from chaos, see John Dixon Hunt, "The Garden as Cultural Object," in *Denatured Visions: Landscape and Culture in the Twentieth Century*, ed. Stuart Wrede and William Howard Adams (New York: Museum of Modern Art, 1991), 19–32, p. 21.
21. On Fra Giovanni Giocondo of Verona and Pacello da Mercogliano see Woodbridge, *Princely Gardens*, 35, 40–42, and Hazlehurst, *Boyceau*, 12–13. Hazlehurst names Girolamo da Napoli rather than Giocondo of Verona as one of the gardeners brought by Charles VIII to France.
22. On Charles's change of heart regarding the design of Amboise see Knecht, *French Renaissance Court*, 143.
23. Du Cerceau's book was entitled *Les plus excellents bastiments de France*; the first volume was published in 1576, the second in 1579. The engravings and text, along with extensive commentary, have recently been republished in *Jacques Androuet du Cerceau: Les Dessins des Plus Excellents Bâtiments de France*, ed. Françoise

Boudon and Claude Mignot (Paris: Picard / Cité de l'architecture et du patrimoine / Le Passage, 2010), where the engravings of Amboise can be found on pages 150–57. Reproductions of du Cerceau's Amboise engravings can also be found in Woodbridge, *Princely Gardens*, 43, and Hazlehurst, *Boyceau*, figures 6, 8.

24. On Mercogliano's garden at Amboise, inspired by Poggio Reale, see Woodbridge, 42, and Knecht, *French Renaissance Court*, 145.
25. On Louis XII's domestic policies see Knecht, *French Renaissance Court*, 2–4. The quote is from page 4.
26. On Machiavelli's assessment of Louis XII see Niccolò Machiavelli, *The Prince*, ed. and trans. David Wootton (Indianapolis, IN: Hackett, 1995), chap. 3, 7–14.
27. On the garden at Blois see Woodbridge, *Princely Gardens*, 42–45, and Knecht, *French Renaissance Court*, 145–49. Dom Antonio de Beatis, who visited in 1517, reported on the gallery.
28. On Francis's single combat with wild boars see Knecht, 90.
29. On Francis's giant court and his incessant travels in his kingdom see Knecht, 32, 39–42.
30. On the dependence of French monarchs in the early Renaissance on the provincial nobility see Robin Briggs, *Early Modern France, 1560–1715*, 2nd ed. (Oxford: Oxford University Press, 1998), 3–6. On French Renaissance kings' neglect of Paris see Knecht, *French Renaissance Court*, 29–31.
31. Quoted in Knecht, 31, 153.
32. On Fontainebleau and its gardens see Knecht, 161–66, and Woodbridge, *Princely Gardens*, 56–61.
33. On Henry II's character and policies see Knecht, *French Renaissance Court*, 17–21.
34. On the divisions in France under Henry II see Briggs, *Early Modern France*, 13–14.
35. On the death of Henry II see Knecht, *French Renaissance Court*, 241–43.

4: THE RETURN OF THE KINGS

1. On the weaknesses of the French monarchy in the Renaissance see Robin Briggs, *Early Modern France, 1560–1715*, 2nd ed. (Oxford: Oxford University Press, 1998), 3–6.
2. On the civil war in France see Briggs, *Early Modern France*, 13–32; on the St. Bartholomew's Day Massacre see pages 21–23.
3. On the Guise clan and its rivals see Briggs, 13–23; on the Catholic League see pages 23–27.
4. On Catherine de Medici see Robert J. Knecht, *The French Renaissance Court, 1483–1589* (New Haven, CT: Yale University Press, 2008), 14–19, 247–79, and Briggs, 15–23.
5. On Catherine de Medici's cultural policies see Knecht, *French Renaissance Court*, chap. 15, 259–79.
6. On the Tuileries see Knecht, 269–71; Kenneth Woodbridge, *Princely Gardens: The Origins and Development of the French Formal Style* (New York: Rizzoli,

1986), 77–79, 112–18; and Michel Baridon, *A History of the Gardens of Versailles*, trans. Adrienne Mason (Philadelphia: University of Pennsylvania Press, 2008), 10–12.

7. On the origins of the Tuileries, de l'Orme, and du Cerceau's engraving, see Knecht, 269–71, and *Jacques Androuet du Cerceau: Les Dessins des Plus Excellents Bâtiments de France*, ed. Françoise Boudon and Claude Mignot (Paris: Picard / Cité de l'architecture et du patrimoine / Le Passage, 2010), 173.
8. Woodbridge, *Princely Gardens*, 80, and *Jacques Androuet du Cerceau*, 175–76.
9. The numbers for Fontainebleau and the Tuileries can vary because some large squares can also be counted as an agglomeration of several smaller ones. Seventeen and thirty-eight are based on a conservative count.
10. Quoted in Baridon, *History of the Gardens of Versailles*, 11.
11. Woodbridge, *Princely Gardens*, 112.
12. The 1573 celebrations at the Tuileries are described in Jean Dorat, *Magnificentissimi Spectaculi, a Regina Regum Matre in hortis suburbanis editi, in Henrici Regis Poloniae invictissimi nuper renunciati gratulationem descriptio* (Paris: Frédéric Morel, royal typographer, 1573). For modern discussions see Woodbridge, *Princely Gardens*, 83; Knecht, *French Renaissance Court*, 265; and Baridon, *History of the Gardens of Versailles*, 11.
13. On Henry IV's war against the League and his conversion see Briggs, *Early Modern France*, 27–29.
14. On the Edict of Nantes see Briggs, 30–31.
15. On the conspiracies of Biron and Bouillon see Yves-Marie Bercé, *The Birth of Absolutism: A History of France, 1598–1661*, trans. Richard Rex (New York: St. Martin's Press, 1992), 11–18.
16. On the challenges facing Henry IV, his assassination, and the succession of 1610 see Bercé, *Birth of Absolutism*, 4–7, 29–41.
17. On the siege and conquest of the Huguenot stronghold of La Rochelle see Bercé, 98–102, and Briggs, *Early Modern France*, 92–94.
18. See Bercé, 137–42.
19. See Baridon, *History of the Gardens of Versailles*, 12.
20. Woodbridge, *Princely Gardens*, 129–33.
21. Quoted in Woodbridge, 121.
22. A royal minority officially ended when the king reached the age of thirteen, but in reality usually lasted much longer. Louis XIII did not gain his independence until he ousted his mother and her favorites in a palace coup when he was sixteen. His son, Louis XIV, became effective ruler only after the death of Cardinal Mazarin in 1661, when the king was twenty-three.
23. On the origins of the Palais de Luxembourg and its gardens, and Marie de Medici's desire to re-create the Pitti Palace and Boboli Gardens, see Woodbridge, *Princely Gardens*, 134–38, and Franklin Hamilton Hazlehurst, *Jacques Boyceau and the French Formal Garden* (Athens: University of Georgia Press, 1966), 49–50.
24. See Baridon, *History of the Gardens of Versailles*, 18.

25. On the château, gardens, and town of Richelieu see Woodbridge, *Princely Gardens*, 143–45, and Baridon, 17–20.
26. The two most famous gardening dynasties, both headquartered in the Tuileries, were the Mollets and the Le Nôtres.
27. On Jacques de Boyceau's life and career see Hazlehurst, *Boyceau*, 2–8.
28. On Boyceau's connection to the Luxembourg gardens and the first gardens of Versailles, see Hazlehurst, *Boyceau*, 48–74. For the parterre designs see Jacques Boyceau, *Traité du Jardinage selon les raisons de la nature et de l'art* (Paris: Michel van Lochom, 1638). All the parterre designs follow the end of the text on page 87. The Luxembourg parterres are on pages 110–24; Versailles, on 127; Saint-Germain-en-Laye, 130–33.
29. Boyceau, *Traité du Jardinage*, 3.
30. Boyceau, 69, also quoted in Hazlehurst, *Boyceau*, 32.
31. Boyceau, 69, and Hazlehurst, 32. Significantly, in 1603 Christoph Grienberger (1564–1636), leading Jesuit mathematician and professor of geometry at the Collegio Romano, made a similar point in a lecture to his colleagues. Nature, he argued, was ruled by mathematical laws, and mathematicians often liberated these elements from their dungeons and "introduced them into the gardens and palaces of kings." Quoted in Volker R. Remmert, "The Art of Garden and Landscape Design and the Mathematical Sciences in the Early Modern Period," in *Gardens, Knowledge, and the Sciences in the Early Modern Period*, ed. Hubertus Fischer, Volker R. Remmert, and Joachim Wolschke-Bulmahn (Basel: Birkhauser, 2016), 9–28, p. 15. Like the kings of France, the Jesuits—and especially the mathematicians among them—were invested in a fixed, hierarchical, and geometrically ordered universe. See Amir Alexander, *Infinitesimal: How a Dangerous Mathematical Theory Shaped the Modern World* (New York: Scientific American / Farrar, Straus and Giroux, 2014).
32. Boyceau, *Traité du Jardinage*, 71; also quoted in Hazlehurst, *Boyceau*, 34.
33. "The triangle being doubled makes a hexagon, an octagon proceeds from a square, and for the pentagon . . . it only remains for it to have the same perfection in gardening as in other works in which it is frequently employed." Boyceau, 71.

5: THE KING'S GARDEN

1. On the king's bedchamber at the center of the palace and the heart of the kingdom see Joël Cornette, "Le Palais du Plus Grand Roi du Monde," in *Versailles. Le pouvoir de la pierre*, ed. Joël Cornette (Paris: Éditions Tallandier, 2006), 27–42, esp. 37.
2. For a powerful account of the experience of approaching the palace of Versailles from the town and emerging into the gardens, see Michel Baridon, *A History of the Gardens of Versailles*, trans. Adrienne Mason (Philadelphia: University of Pennsylvania Press, 2008), 1–2.
3. When Alain Manesson Mallet, master mathematician at Louis XIV's court, published a book on practical geometry in 1702, the majority of the engravings were

of Versailles. See Alain Manesson Mallet, *La Géométrie Pratique Divisée en Quatre Livres* (Paris: Anisson, Imprimerie Royale, 1702).

4. The estimate of the area of the Versailles gardens is from Baridon, *History of the Gardens of Versailles*, 112, quoting the historian Thierry Mariage. The estimate of the number of workers at Versailles provided in 1685 by the courtier Dangeau is from Cornette, "Le Palais du Plus Grand Roi," 32.
5. The anecdote is from W. H. Lewis, *The Splendid Century: Life in the France of Louis XIV* (New York: Morrow Quill, 1978; orig. pub. 1953), 4.
6. On Louis XIV's succession, and the queen mother's circumvention of her husband's will, see Yves-Marie Bercé, *The Birth of Absolutism: A History of France, 1598–1661*, trans. Richard Rex (New York: St. Martin's Press, 1992), 157–58.
7. On Louis XIV's upbringing under Mazarin see Lewis, *Splendid Century*, chap. 1.
8. See Louis XIV, *Mémoires for the Instruction of the Dauphin*, trans. Paul Sonnino (New York: Free Press, 1970), 24.
9. For an account of the Fronde and its causes see Bercé, *Birth of Absolutism*, 164–78.
10. On Louis XIV's humiliation during the Fronde see Lewis, *Splendid Century*, 7–8. The scene of the young king feigning sleep as a delegation of Frondeurs enters his room is imaginatively described in Alexandre Dumas's musketeers novel *Twenty Years After.*
11. Bercé, *Birth of Absolutism*, 177–79, and Kenneth Woodbridge, *Princely Gardens: The Origins and Development of the French Formal Style* (New York: Rizzoli, 1986), 181.
12. On Mazarin's death and Louis XIV's assertion of personal rule see Lewis, *Splendid Century*, 8–9, and Pierre Goubert, *Louis XIV and Twenty Million Frenchmen*, trans. Anne Carter (New York: Pantheon Books, 1970), 19.
13. See Louis XIV, *Mémoires*, 31.
14. On Louis XIV's suppression of dissent see Robin Briggs, *Early Modern France, 1560–1715*, 2nd ed. (Oxford: Oxford University Press, 1998), 138–40.
15. On Louis XIV's reform of the army and navy see Briggs, *Early Modern France*, 67, 69, 141. On Louis XIV's early wars see Goubert, *Louis XIV*, 98–114, and Briggs, 142–45.
16. On Louis XIV's revamping of the royal palaces and building projects in Paris see Goubert, 80–81. On the *carrousel* see Peter Burke, *The Fabrication of Louis XIV* (New Haven, CT: Yale University Press, 1992), 66–67.
17. Paul Pellisson, "Panegyric of Louis XIV," in *The Greatness of Louis XIV: Myth or Reality?*, ed. William F. Church, 2nd ed. (Lexington, MA: D.C. Heath, 1972), 3–8, p. 4.
18. Louis XIV, *Mémoires.*
19. Louis XIV, 43.
20. Louis XIV, 155.
21. Ibid.
22. Louis XIV, *Mémoires*, 43, 130.
23. Louis XIV, 68.
24. Louis XIV, 64.
25. Louis XIV, 188.

26. For the deep interconnections between geometry and faith in a rigid, hierarchical, and unalterable political order in the early modern world see Amir Alexander, *Infinitesimal: How a Dangerous Mathematical Theory Shaped the Modern World* (New York: Scientific American / Farrar, Straus and Giroux, 2014).
27. Quoted in Baridon, *History of the Gardens of Versailles*, 28.
28. Quoted in Baridon, 25, 28.
29. Quoted in Baridon, 28.
30. See discussion in Alexander, *Infinitesimal*, 216–18. For the quote see Thomas Hobbes, *Leviathan*, ed. Edwin Curley (Indianapolis: Hackett Publishing, 1994), 76.
31. On Hobbes and Sorbière see Alexander, *Infinitesimal*, 3–8.
32. Jacques-Bénigne Bossuet, *Politics Drawn from the Very Words of Holy Scripture*, trans. and ed. Patrick Riley (Cambridge, UK: Cambridge University Press, 1990), first published in 1709 as *Politique tirée des propres paroles de l'Écriture Sainte.*
33. Quoted in Baridon, *History of the Gardens of Versailles*, 26–27.
34. For a discussion of Bossuet's teachings and his views on Hobbes see Patrick Riley, introduction to Bossuet, *Politics*, xiii–lxviii.
35. On the order of precedence at court, and the related anecdotes, see Lewis, *Splendid Century*, 41–47.
36. Baridon, *History of the Gardens of Versailles*, 44.
37. Louis XIV, *Mémoires*, 144.
38. The discussion of the emergence of ballet in the courts of the kings of France in the sixteenth and seventeenth centuries is based on "Kings of Dance," chapter 1 in Jennifer Homans, *Apollo's Angels: A History of Ballet* (New York: Random House, 2010), 3–48.
39. On Beaujoyeulx and the *Ballet comique de la Reine* of 1581 see Homans, *Apollo's Angels*, 7–8.
40. Homans, 23–26.
41. Homans, 19–23.
42. Bernard Le Bovier de Fontenelle, *Histoire du renouvellement de l'Académie royale des sciences en mdcxcix et les éloges historiques* (Amsterdam: Pierre du Coup, 1719–20), vol. 1, 14, quoted in J. L. Heilbron, "Introductory Essay," in *The Quantifying Spirit in the Eighteenth Century*, ed. Tore Frängsmyr, J. L. Heilbron, and Robin E. Rider (Berkeley: University of California Press, 1990), 1–25, p. 1.
43. On Louis XIII's Versailles see F. Hamilton Hazlehurst, *Gardens of Illusion: The Genius of André Le Nostre* (Nashville, TN: Vanderbilt University Press, 1980), 59–61, and Robert W. Berger and Thomas F. Hedin, *Diplomatic Tours in the Gardens of Versailles under Louis XIV* (Philadelphia: University of Pennsylvania Press, 2008), 5.
44. On Le Nôtre's life see Hazlehurst, *Gardens of Illusion*, 1–8; Woodbridge, *Princely Gardens*, 185–88; and Baridon, *History of the Gardens of Versailles*, 127–38. The quote by Saint-Simon is from Hazlehurst, 6.
45. The story is told in Hazlehurst, 7.

46. Ibid.
47. On the creation of the gardens of Vaux-le-Vicomte see Woodbridge, *Princely Gardens*, 188–95, and Hazlehurst, *Gardens of Illusion*, 17–45.
48. This is known from a plan of Vaux-le-Vicomte by the engraver Israël Silvestre. The pattern came to be known as "crow's feet," or *patte d'oie* in French. See Hazlehurst, 22, plate 6.
49. André Mollet, *Le Jardin de Plaisir, contenant plusieurs desseins de jardinage . . .* (Stockholm: Henry Kayser, 1651), 31.
50. Louis XIV, *Mémoires*, 43.
51. Louis XIV, 144.
52. Ibid.
53. Even the potager du roi served ideological as well as practical purposes. See Chandra Mukerji, "The Power of the Sun King at the Potager du Roi," in *Gardens, Knowledge, and the Sciences in the Early Modern Period*, ed. Hubertus Fischer, Volker R. Remmert, and Joachim Wolschke-Bulmahn (Basel: Birkhauser, 2016), 55–74.
54. For a modern comparison of the gardens at Vaux and Versailles, which strongly favor the former, see Claire Goldstein, *Vaux and Versailles: The Appropriations, Erasures, and Accidents That Made Modern France* (Philadelphia: University of Pennsylvania Press, 2007), especially 116–18. A critique along similar lines can be found in Allen S. Weiss, *Mirrors of Infinity: The French Formal Gardens and 17th-Century Metaphysics* (New York: Princeton Architectural Press, 1995), 46–47.
55. On Félibien and Scudéry at Versailles see Goldstein, *Vaux and Versailles*. For some of their original texts see André Félibien, *Les Fêtes de Versailles: Chroniques de 1668 & 1674* (Paris: Éditions Dédales, 1994), and Madeleine de Scudéry, *La Promenade de Versailles* (Paris: Barbin, 1669; repr. Geneva: Slatkine, 1979).
56. On Martin's comment see Chandra Mukerji, *Territorial Ambitions and the Gardens of Versailles* (Cambridge, UK: Cambridge University Press, 1997), 213–14. On the Russian ambassador's comments see Berger and Hedin, *Diplomatic Tours*, 26.
57. Bayle St. John, ed. and trans., *The Memoirs of the Duke of Saint-Simon on the Reign of Louis XIV and the Regency* (London: Swan Sonnenschein, 1889), vol. II, 369.
58. Woodbridge, *Princely Gardens*, 223.
59. On the Doge of Genoa's tour of Versailles see Berger and Hedin, *Diplomatic Tours*, 28–30.
60. Known as the Du Bus plan and dating from around 1661, the map is reproduced in Hazlehurst, *Gardens of Illusion*, 61.
61. The following account of Versailles in the early 1660s follows François de la Pointe's plan of the gardens dating from around 1665. See Hazlehurst, 72.
62. On the Pleasures of the Enchanted Isle, see François Bluche, *Louis XIV*, trans. Mark Greengrass (New York: Franklin Watts, 1990), 179–83.
63. Technically Fouquet and Colbert held different positions, though in practice both were responsible for the kingdom's finances. Fouquet's position of superintendent

of finances was abolished after his banishment, and Colbert held the title of controller general of finances.

64. Quoted in Woodbridge, *Princely Gardens*, 202.
65. On the expansions at Versailles at this time see Woodbridge, 197–206, and Hazlehurst, *Gardens of Illusion*, 73–84.
66. On the Cassinis and the mapping of France see Baridon, *History of the Gardens of Versailles*, 82–83; Ken Alder, *The Measure of All Things: The Seven-Year Odyssey and Hidden Error that Transformed the World* (New York: Free Press, 2002), 14–19; and Mary Terrall, *The Man Who Flattened the Earth: Maupertuis and the Sciences in the Enlightenment* (Chicago: University of Chicago Press, 2002), 89–91.
67. Jean-Dominique Cassini, *Carte de France Corrigée par ordre du Roy sur les observations de Mss. de l'Academie des Sciences* (Paris, 1693).
68. On the connection between Colbert's ambition to turn France into a maritime state and the fleet on the Grand Canal see Baridon, *History of the Gardens of Versailles*, 53. Locke's account is from John Locke, entry of June 23, 1677, *Locke's Travels in France 1675–1679*, ed. John Lough (Cambridge, UK: Cambridge University Press, 1953), 152.
69. On the menagerie at Versailles see Baridon, 54.
70. For a brilliant analysis of Versailles and the meanings it encodes see Mukerji, *Territorial Ambitions*, especially chapter 6, 248–99. On cascades, staircases, and the Slave Fountain see 285–89.
71. Jacques-Bénigne Bossuet, "The Divine Right of Kings" (excerpts from the *Politique*), in Church, ed., *The Greatness of Louis XIV*, 11.

6: BEYOND VERSAILLES

1. For Louis XIV's guide to the gardens of Versailles see Louis XIV, *Manière de montrer les Jardins de Versailles*, ed. Simone Hoog (Paris: Éditions de la Réunion des Musées Nationaux, 1982).
2. Robert W. Berger and Thomas F. Hedin, *Diplomatic Tours in the Gardens of Versailles under Louis XIV* (Philadelphia: University of Pennsylvania Press, 2008), 26.
3. Berger and Hedin, *Diplomatic Tours*, 28.
4. Berger and Hedin, 39.
5. On Le Nôtre's designs for Greenwich Park see F. Hamilton Hazlehurst, *Gardens of Illusion: The Genius of André Le Nostre* (Nashville, TN: Vanderbilt University Press, 1980), 3, 377–79.
6. Voltaire, *Le Siècle de Louis XIV*, ed. René Groos (Paris: Librairie Garnier Frères, 1929), vol. 1, 3. Translation follows William F. Church, ed., *The Greatness of Louis XIV: Myth or Reality?*, 2nd ed. (Lexington, MA: D.C. Heath, 1972), 55.
7. Voltaire, *Le Siècle de Louis XIV*, quoted in Michel Baridon, *A History of the Gardens of Versailles*, trans. Adrienne Mason (Philadelphia: University of Pennsylvania Press, 2008), 23–24.
8. On the trivium in the Piazza del Popolo and elsewhere in Renaissance Rome see

Spiro Kostof, *The City Shaped: Urban Patterns and Meanings Through History* (Boston: Bulfinch Press, 1991), 235–36.

9. On Sixtus V and Fontana's urban plan for Rome see Kostof, *The City Shaped*, 218, 242–43.
10. On the proposals for rebuilding fire-scorched London see John W. Reps, *The Making of Urban America: A History of City Planning in the United States* (Princeton: Princeton University Press, 1965), 15–19, 163–64.
11. The book was ultimately edited and published only in 2000. See John Evelyn, *Elysium Britannicum, or the Royal Gardens*, ed. John E. Ingram (Philadelphia: University of Pennsylvania Press, 2000). On Evelyn's lifelong quest to complete the work see Michael Leslie, "'Without Design, Fate, or Force': Why Couldn't John Evelyn Complete the *Elysium Britannicum*?," in *Gardens, Knowledge, and the Sciences in the Early Modern Period*, ed. Hubertus Fischer, Volker R. Remmert, and Joachim Wolschke-Bulmahn (Basel: Birkhauser, 2016), 29–54.
12. Quoted in Volker R. Remmert, "The Art of Garden and Landscape Design and the Mathematical Sciences in the Early Modern Period," in Fischer et al., *Gardens, Knowledge, and the Sciences*, 9–28, p. 21.
13. On the St. Petersburg trivium and the influence of Versailles see Kostof, *The City Shaped*, 236, 239, 252.
14. On Vienna's Ringstrasse as a representation of the vision of the liberal bourgeoisie that dominated the imperial and city government in the second half of the nineteenth century, see Carl E. Schorske, *Fin-de-Siecle Vienna: Politics and Culture* (New York: Alfred A. Knopf, 1980), chapter 2: "The Ringstrasse, Its Critics, and the Birth of Urban Modernism," 24–115. By the eve of World War I, according to Schorske, the Ringstrasse and all it stood for had come under attack from both left and right.
15. On Haussmann's Paris and its echoes of Versailles see Vincent Scully, "Architecture: The Natural and the Manmade," in *Denatured Visions: Landscape and Culture in the Twentieth Century*, ed. Stuart Wrede and William Howard Adams (New York: Museum of Modern Art, 1991), 16–17.
16. In 1975, following the fall of South Vietnam, Saigon was renamed Ho Chi Minh City, and remains so today.
17. On the rue de la Paix in Paris as a model for the rue Catinat in Saigon see Walter E. J. Tips, introduction to Pierre Barrelon, Brossard de Corbigny, Charles Lemire, and Gaston Cahen, *Cities of Nineteenth Century Colonial Vietnam: Hanoi, Saigon, Hue, and the Champa Ruins*, trans. Walter E. J. Tips (Bangkok: White Lotus Press, 1999), ix–xii, p. ix. The quote is from Pierre Barrelon, "Saigon," 31–95, p. 59.
18. A description of the Boulevard Norodom in the late nineteenth century can be found in Pierre Barrelon, *Saigon* (Ithaca, NY: Cornell University Library, 1893), 55–56.
19. On the new capital of India, the motives for its creation, and its design, construction, meaning, and reception, see David A. Johnson, *New Delhi: The Last Imperial City* (New York: Palgrave Macmillan, 2015). On New Delhi as a site of colonial coercion

and control see Stephen Legg, *Spaces of Colonialism: Delhi's Urban Governmentalities* (Malden, MA: Blackwell Publishing, 2007).

20. Baker's quotation of Wren and his views on the architecture of the new capital came from his editorial entitled "The New Delhi," published in *The Times*, October 3, 1912.
21. Hardinge wrote of his wish to copy Versailles in a 1912 letter, quoted in Johnson, 140.
22. Known today as Rashtrapati Bhavan, it serves as the official residence of the president of India.
23. Quoted in Thomas R. Metcalf, "Architecture and the Representation of Empire: India, 1860–1910," *Representations* 6 (Spring 1984): 37–65, p. 61. In the end, despite his protestations to the contrary, the "silly Mughal-Hindu stuff" made significant inroads into Lutyens's design, manifested most prominently in the outlines of the grand dome.
24. King's Way, or Kingsway, is known today as Rajpath Marg.
25. To determine the proper dimensions of the boulevard Hardinge requested the precise measurements of Paris's Champs-Élysées, Berlin's Unter den Linden, and Vienna's Ringstrasse. See Johnson, *New Delhi*, 137–38.
26. Queen's Way is today Janpath Road.
27. On Hardinge's imperial vision and the politics that shaped New Delhi see Johnson, *New Delhi*, 21–38, 183–95.
28. Johnson, 125.
29. Quoted in Johnson, 144.
30. Quoted in Johnson, 191–92.
31. The Indian Legislative Assembly, created in the wake of the "Government of India Act" of 1919, was also located near the eastern end of King's Way, not far from the Viceroy's Palace and Secretariats. It was, however, awkwardly positioned—off the main boulevard, on lower ground, and to the side. This was an eloquent architectural expression of the ambiguous position of the Indian Legislative Assembly in the Raj: it was an awkward late addition with little real power, and clearly not an essential branch of government.
32. On Burnham and his career see Christopher Vernon, "Daniel Hudson Burnham and the American City Imperial," *Thesis Eleven* 123, no. 1 (2014): 80–105. For more on Pierre L'Enfant's plan for the federal capital see chapter 7.
33. On Burnham's recruitment by Forbes and his travels in the Philippines see Thomas S. Hines, "The Imperial Façade: Daniel H. Burnham and American Architectural Planning in the Philippines," *Pacific Historical Review* 41, no. 1 (February 1972), 33–53, pp. 38–43. The quote is from page 43.
34. Burnham to Charles Moore, quoted in Hines, "Imperial Façade," 43.
35. Quoted in Vernon, "Daniel Hudson Burnham," 89.
36. See Vernon, 90.
37. Quoted in Hines, "Imperial Façade," 44.
38. As Herbert Baker, co-designer of New Delhi, explained in a *Times* editorial, only Classical architectural styles should be used in British India, because they alone

"embody the idea of law and order which has been produced out of chaos by the British Administration." See Baker, "The New Delhi," *The Times*, October 3, 1912.

7: THE EUCLIDEAN REPUBLIC

1. See Elizabeth S. Kite, *L'Enfant and Washington, 1791–1792* (Baltimore: Johns Hopkins University Press, 1929), 47. For summaries of the entire affair, including quotes, see Reps, *Making of Urban America*, 240–56, and Spiro Kostof, *The City Shaped: Urban Patterns and Meanings Through History* (Boston: Bulfinch Press, 1991), 209–11.
2. For L'Enfant's letter to Washington, dated September 11, 1789, offering his services in designing the capital, see Kite, *L'Enfant and Washington*, 34.
3. See Joseph J. Ellis, *Founding Brothers: The Revolutionary Generation* (New York: Alfred A. Knopf, 2000), chap. 2, 69–71.
4. Reps, *Making of Urban America*, 240–41, and Ellis, *Founding Brothers*, 69–71.
5. Ellis, 71.
6. Ibid.
7. For Jefferson's account of his encounter with Hamilton and the discussions and compromise that followed see *The Papers of Thomas Jefferson*, ed. Julian P. Boyd, vol. 17 (Princeton: Princeton University Press, 1965), 205–208.
8. For a beautiful account of Jefferson's intervention in the fight over the national capital and the compromise he brokered, see Ellis, *Founding Brothers*, chap. 2, 48–80.
9. For "Speculating phalanx" see *Papers of Thomas Jefferson*, vol. 17, 207.
10. *Papers of Thomas Jefferson*, vol. 17, 206.
11. While Jefferson's account is essentially accurate, the actual compromise was more complex and involved a series of negotiations over several weeks. For a full account see Ellis, *Founding Brothers*, chap. 2.
12. For correspondence and documents pertaining to Jefferson's involvement in the design and construction of the city of Washington see Saul K. Padover, ed., *Thomas Jefferson and the National Capital* (Washington: U.S. Government Printing Office, 1946). For an overview of Jefferson's role see C. M. Harris, "Washington's Gamble, L'Enfant's Dream: Politics, Design, and the Founding of the National Capital," *The William and Mary Quarterly* 56, no. 3 (July 1999), 527–64.
13. See Kite, *L'Enfant and Washington*, 47–48.
14. Jefferson to L'Enfant, April 10, 1791, in Padover, *Jefferson and the National Capital*, 59. Also in Kite, *L'Enfant and Washington*, 48–50.
15. Quoted in J. J. Jusserand, introduction to Kite, *L'Enfant and Washington*, 24.
16. On the rise and fall of L'Enfant see Reps, *Making of Urban America*, 240–56. On his relationship with Jefferson, and Jefferson's role in his fall, see Harris, "Washington's Gamble, L'Enfant's Dream," especially 544–50.
17. On L'Enfant's background and career before sailing to America see Jusserand, introduction, 2–3, and Iris Miller, *Washington in Maps, 1606–2000* (New York: Rizzoli, 2002), 17–21.

18. See Jusserand, 2–4.
19. Jusserand, 4–5, 9–11; on the French examples that helped shape L'Enfant's design for Washington, see Miller, *Washington in Maps*, 12–21.
20. On the development of L'Enfant's plan see Reps, *Making of Urban America*, 248–52, and Miller, 34–40.
21. On the size of American cities see Howard P. Chudacoff and Judith E. Smith, *The Evolution of American Urban Society*, 5th ed. (Upper Saddle River, NJ: Prentice Hall, 2000), especially the tables on pages 7 and 55.
22. See Kite, *L'Enfant and Washington*, 47.
23. See L'Enfant's report to Washington, June 1791, in Kite, 52–58; the quote is on 55.
24. The final master plan for the city was created not by L'Enfant but by his successor, Andrew Ellicott, who made subtle changes to the Frenchman's original design. Most notably, Massachusetts Avenue, which L'Enfant's plan angles gradually to the southeast, becomes arrow-straight in Ellicott's version. The changes outraged L'Enfant, who considered his plan "unmercifully spoiled." See Miller, *Washington in Maps*, 44–47.
25. See the comments to L'Enfant's plan as presented in Philadelphia in August of 1791 in Kite, *L'Enfant and Washington*, 62–66.
26. Kite, 45.
27. George Bancroft, *History of the Formation of the Constitution of the United States of America*, vol. 2 (New York: D. Appleton, 1882), 284. Quoted in Joseph J. Ellis, *The Quartet: Orchestrating the Second American Revolution, 1783–1789* (New York: Alfred A. Knopf, 2015), 141.
28. In his correspondence L'Enfant refers repeatedly to the United States as an "Empire." See, for instance, his 1789 letter to Washington offering his services as the capital's architect—Kite, *L'Enfant and Washington*, 34—and his report to Washington from January 1792: Kite, 110–16, p. 116.
29. Reps, *Making of Urban America*, 256.
30. Quoted in Harris, "Washington's Gamble, L'Enfant's Dream," 552.
31. For a full account of Jefferson's designs for the capital city see Reps, *Making of Urban America*, 245–48.
32. Jefferson, "Proceedings to be had under the Residence Act," November 29, 1790, and "Jefferson's Report to Washington on Meeting Held at Georgetown, September 14, 1790," in Padover, *Jefferson and the National Capital*, 30–36. Quotes are from pages 31, 34.
33. For the size of Philadelphia in 1790 see Reps, *Making of Urban America*, 246.
34. Jefferson, "Report to Washington," in Padover, *Jefferson and the National Capital*, 31, 35.
35. Jefferson's practice of collecting city plans was so well known that L'Enfant himself asked to borrow this rare collection to aid in his design of the American capital, and Jefferson obligingly sent him plans of "Frankfort on the Mayne, Carlsruhe, Amsterdam, Strasburg, Paris, Orleans, Bordeaux, Lyon, Montpellier, Marseilles, Turin, and Milan." See Jefferson to Washington, April 10, 1791, in Padover, 60–61, p. 60.
36. See Jefferson, "Proceedings," in Padover, 30–36, pp. 31–32.

37. On Jefferson's view that new laws must be rewritten by each successive generation, see Jefferson to James Madison, September 6, 1789, in Thomas Jefferson, *Writings*, ed. Merrill D. Peterson (New York: Library of America; 1984; repr. 2011), 959–64.
38. A second "father" was undoubtedly Daniel H. Burnham, who was chiefly responsible for the revival of L'Enfant's design at the turn of the twentieth century. In the nineteenth century L'Enfant's "grand boulevard" leading up to the Capitol was recast as a naturalistic English-style garden, with dense greenery and winding paths, similar in concept to Central Park in New York. L'Enfant's plan was revived in 1901 by the U.S. Senate Park Commission chaired by Burnham, which designed the National Mall and the city as we know them today. Implementation of Burnham's plan (also known as the McMillan Plan for the senator who organized the commission) took decades, and was not completed until 1934.
39. On Washington's continued involvement in the design and construction of the capital see Harris, "Washington's Gamble, L'Enfant's Dream."
40. For example, in his final years Washington hatched a plan to have his tomb placed under the dome of the Capitol, thereby making it the symbolic heart of the nation. This despite the fact that he personally preferred to be buried on his estate at Mount Vernon, alongside his wife, Martha—as indeed he eventually was. The plan, and Washington's motives for it, are discussed in Harris.

CONCLUSION

1. On Bolyai and his discovery of non-Euclidean geometry see Amir Alexander, *Duel at Dawn: Heroes, Martyrs, and the Rise of Modern Mathematics* (Cambridge, MA: Harvard University Press, 2010), 215–51; Jeremy J. Gray, *János Bolyai, Non-Euclidean Geometry, and the Nature of Space*, Publications of the Burndy Library (Cambridge, MA: MIT Press, 2004); and Roberto Bonola, *Non-Euclidean Geometry: A Critical and Historical Study of Its Development*, trans. H. S. Carslaw (Mineola, NY: Dover Publications, 1955).
2. The proof that the sum of the angles of a triangle is equal to two right angles, for example (see the introduction, pages 13–14), depended on transferring two of the triangle's interior angles to locations adjoining the third by drawing a parallel.
3. Nikolai Lobachevskii (1792–1856), a professor of mathematics at the University of Kazan, Russia, developed remarkably similar ideas at roughly the same time as Bolyai, and the two are often mentioned as the co-discoverers of non-Euclidean geometry. Others, including Carl Friedrich Gauss, are also credited in some accounts as discoverers, or at least as major contributors to the discovery.
4. In the 1840s Carl Friedrich Gauss identified the different values of k as corresponding to what he called different "curvatures" of space. This implied that spaces with differing curvatures have different geometries. All the geometries described by Bolyai apply to spaces of constant negative curvature, whereas Euclidean space is "flat," i.e., of zero curvature. In his 1854 *Habilitationsschrift*, or dissertation, Gauss's student Bernhard Riemann expanded the concept of non-Euclidean geometries to any higher dimension, as well as to spaces of varying curvature. See Gray, *János Bolyai*, and Bonola, *Non-Euclidean Geometry*. I thank Professor

Jeremy Gray for his help in clarifying the sequence of discoveries and their significance.

5. On the struggle between the advocates of infinitesimal methods and their opponents, and the social visions embodied by each, see Amir Alexander, *Infinitesimal: How a Dangerous Mathematical Theory Shaped the Modern World* (New York: Scientific American / Farrar, Straus and Giroux, 2014).
6. See Amir Alexander, "The Estrangement of the American Landscape," *Social Research* 85, no. 2 (Summer 2018), 323–50.

ACKNOWLEDGMENTS

I would like to thank Mary Terrall, Ted Porter, and Soraya de Chadarevian, my colleagues in the history of science field at UCLA, for their friendship and support throughout the writing of this book. Margaret "Peg" Jacob provided friendship and much valuable advice, and Diane Mizrachi helped preserve my spirits through the ups and downs of a long project. Many thanks as well to J. B. Shank, Ofer Gal, Massimo Mazzotti, Raz Chen-Morris, Kevin Lambert, Jennifer Nelson, and Mario Biagioli for many conversations about the mathematical culture of early modern Europe. These found their way into this book in more ways than I can enumerate, or even fully know.

I am grateful to Amanda Moon, formerly of Farrar, Straus and Giroux, for acquiring the manuscript and seeing it through the early stages of the editing process. When she departed, Laird Gallagher took over, providing detailed and insightful comments on the entire manuscript. Julia Ringo saw the book through the production process, which was managed and carried out by Nancy Elgin, Peter Richardson, Debra Helfand, and Annie Gottlieb. Steve Weil did a magnificent job promoting the book, and Eric Chinski oversaw the project from beginning to end. I thank them all.

None of my writing ventures would have come to fruition without the sound advice and warm encouragement of my wonderful agent, Lisa Adams. Nor would they be possible without my lifelong friend Daniel Baraz. Even as life took us to opposite sides of the world, the friendship endured and has sustained me through four and a half decades (so far).

Bonnie, my love and my wife of twenty-eight years, is an artist and a teacher. Her beautiful creations surprise and delight me, her kindness and generosity inspire me. It goes without saying that not a word of this book could have been written without her love and support. Our children, Jordan and Ella, are adults now, and embarking on their own independent life paths. I thank them for reminding me that there are sometimes more important things than finishing one more paragraph.

My mother, Esther Alexander, was a fighter. As a fifteen-year-old girl in Nazi-occupied Budapest she joined the resistance, delivering suitcases filled with forged documents to drop points across the embattled city. Captured and imprisoned, she was saved by an American bombing raid that partially destroyed the police headquarters where she was being held. She climbed out of the rubble and hid nearby as machine-gun fire rattled through the dawn, killing all who were left behind.

I cannot hope to emulate my mother's heroism. But I learned from her that some things are worth fighting for, no matter the odds and no matter the risks. I dedicate this book to her.

INDEX

ILLUSTRATION CREDITS

Page 4 (Figure 1): © RMN-Grand Palais / Art Resource, NY

Page 7 (Figure 2): Wikimedia Commons, https://commons.wikimedia.org/wiki/File:Versailles_Plan_Jean_Delagrive.jpg

Page 13 (Figure 3): *Euclid's Elements: All Thirteen Books in One Volume*, ed. Dana Densmore, trans. Thomas L. Heath (Santa Fe, NM: Green Lion Press, 2002).

Page 38 (Figure 4): Reproduced from Samuel Y. Edgerton, Jr., *The Renaissance Rediscovery of Linear Perspective* (New York: Basic, 1975), 126

Page 42 (Figure 5): Wikimedia Commons, https://commons.wikimedia.org/wiki/File:Masaccio,_trinit%C3%A0.jpg

Page 43 (Figure 6): Wikimedia Commons, https://commons.wikimedia.org/wiki/File:Masaccio7.jpg

Page 44 (Figure 7): Wikimedia Commons, https://commons.wikimedia.org/wiki/File:Don_Lorenzo_Monaco_002.2.jpg

Page 52 (Figure 8): Reproduced from Leon Battista Alberti, *On Painting*, trans. Cecil Grayson (London: Penguin, 1999), 55.

Page 53 (Figure 9): Reproduced from Edgerton, *The Renaissance Rediscovery of Linear Perspective*, 45. Edgerton combines figures 8 and 10 from Alberti's *On Painting*, 55, 57.

Page 66 (Figure 10): Alinari / Art Resource, NY

Page 67 (Figure 11): bpk Bildagentur / Kupferstichkabinett / Art Resource, NY

Page 89 (Figure 12): Reproduced from *Jacques Androuet du Cerceau: Les Dessins des Plus Excellents Bâtiments de France*, ed. Françoise Boudon and Claude Mignot (Paris: Picard / Cité de l'architecture et du patrimoine / Le Passage, 2010), 155.

Page 96 (Figure 13): Reproduced from *Jacques Androuet du Cerceau*, 161.

Page 109 (Figure 14): Reproduced from *Jacques Androuet du Cerceau*, 175.

Page 123 (Figure 15): Reproduced from Kenneth Woodbridge, *Princely Gardens: The Origins and Development of the French Formal Style* (New York: Rizzoli International Publications, 1986), 145.

Page 133 (Figure 16): Wikimedia Commons, https://commons.wikimedia.org/wiki/File:Plan_g%C3%A9n%C3%A9ral_de_Versailles,_son_parc,_son_Louvre,_ses_jardins,_ses_fontaines,_ses_bosquets_et_sa_ville_par_N_de_Fer_1700_-_Gallica_2012.jpg

Page 163 (Figure 17): Esther Westerveld, Wikimedia Commons, https://commons.wikimedia.org/wiki/File:Kasteel_van_Vaux-le-Vicomte_-_Maincy_06.jpg

Page 175 (Figure 18): Reproduced from Franklin Hamilton Hazlehurst, *Jacques Boyceau and the French Formal Garden* (Athens, GA: University of Georgia Press, 1966), 86, plate 57.

Page 177 (Figure 19): Produced by César-François Cassini de Thury (Cassini III) for the Royal Academy of Sciences, Paris, 1744.

Page 181 (Figure 20): Reproduced from Alain Manesson Mallet, *La Géométrie Pratique Divisée en Quatre Livres* (Paris: Anisson, Imprimerie Royale, 1702), book 1, 63, plate XXX.

Page 199 (Figure 21): Reproduced from Spiro Kostof, *The City Shaped: Urban Patterns and Meanings Through History* (Boston: Bulfinch Press, 1991), 242–43, plate 21.

Page 200 (Figure 22): Reproduced from John W. Reps, *The Making of Urban America: A History of City Planning in the United States* (Princeton: Princeton University Press, 1965), 17, figure 9.

Page 200 (Figure 23): Reproduced from John W. Reps, *The Making of Urban America: A History of City Planning in the United States* (Princeton: Princeton University Press, 1965), 16, figure 8.

Page 205 (Figure 24): Dorling Kindersley Ltd. / Alamy Stock Photo

Page 212 (Figure 25): Public domain (author scan)

Page 215 (Figure 26): Public domain, https://architexturez.net/file/britannica1910-jpg

Page 217 (Figure 27): Sueddeutsche Zeitung Photo / Alamy Stock Photo

Page 222 (Figure 28): Courtesy of the Newberry Library

Page 228 (Figures 29 and 30): Reproduced from John W. Reps, *The Making of Urban America: A History of City Planning in the United States* (Princeton: Princeton University Press, 1965), 246–47.

Page 239 (Figure 31): Library of Congress, Geography and Map Division

Page 242 (Figure 32): Library of Congress, Prints and Photographs Division, Carol M. Highsmith Archive

Page 261 (Figure 33): Wikimedia Commons, https://commons.wikimedia.org/wiki/File:Parallel_postulate_en.svg

A NOTE ABOUT THE AUTHOR

Amir Alexander teaches history at the University of California–Los Angeles. He is the author of *Infinitesimal*, *Duel at Dawn*, and *Geometrical Landscapes*. His writing has appeared in *The New York Times* and the *Los Angeles Times*, and his work has been featured in *Nature* and *The Guardian*, on NPR, and elsewhere. He lives in Los Angeles, California.